MW00680065

Biotechnology
Intelligence
Unit

Policy Controversy in Biotechnology: An Insider's View

Henry I. Miller, M.D.

Hoover Institution
Stanford University
Stanford, California, U.S.A.

Academic Press

R.G. Landes Company
Austin

Biotechnology Intelligence Unit

POLICY CONTROVERSY IN BIOTECHNOLOGY: AN INSIDER'S VIEW

R.G. LANDES COMPANY
Austin, Texas, U.S.A.

This book is printed on acid-free paper.

Please address all inquiries to the Publisher:
R.G. Landes Company
810 S. Church Street, Georgetown, Texas, U.S.A. 78626
Phone: 512/ 863 7762; FAX: 512/ 863 0081

Academic Press, Inc.
525 B Street, Suite 1900, San Diego, California, U.S.A. 92101-4495

United Kingdom Edition published by Academic Press Limited
24-28 Oval Road, London NW1 7DX, United Kingdom

Library of Congress Catalog Number: 660'.6--dc20
International Standard Book Number (ISBN): 0-12-496725-6

Printed in the United States of America

Library of Congress Cataloging-in-Publication Data

Miller, Henry I.
Policy controversy in biotechnology : an insider's view / Henry I. Miller.
p. cm. -- (Biotechnology intelligence unit)
Includes bibliographical references and index.
ISBN 1-57059-408-2 (alk. paper)
1. Biotechnology--Government policy--United States. I. Title. II. Series.
TP248.185.M55 1996
660'.6--dc20

96-35982
CIP

PUBLISHER'S NOTE

R.G. Landes Company publishes six book series: *Medical Intelligence Unit, Molecular Biology Intelligence Unit, Neuroscience Intelligence Unit, Tissue Engineering Intelligence Unit, Biotechnology Intelligence Unit* and *Environmental Intelligence Unit.* The authors of our books are acknowledged leaders in their fields and the topics are unique. Almost without exception, no other similar books exist on these topics.

Our goal is to publish books in important and rapidly changing areas of bioscience and environment for sophisticated researchers and clinicians. To achieve this goal, we have accelerated our publishing program to conform to the fast pace in which information grows in bioscience. Most of our books are published within 90 to 120 days of receipt of the manuscript. We would like to thank our readers for their continuing interest and welcome any comments or suggestions they may have for future books.

Shyamali Ghosh
Publications Director
R.G. Landes Company

Dedication

To my parents.

CONTENTS

FOREWORD

Prior to 1900, there was no direct regulation of consumer products by the federal government in the United States. Congress had investigated the adulteration and misbranding of food and drugs for the last two decades of the 19th century but had stopped short of enacting regulatory legislation because of a deep conviction that this was a matter for state and local governments. In the first decade of the 20th century, however, Congress embarked upon a series of legislative enactments that initiated what has become a century of increasingly more comprehensive and intrusive regulation of consumer products found in the American marketplace. Today, there is no consumer product that is not subject to at least one, and in many instances several, federal regulatory statutes and a bewildering array of implementing regulations, guidelines and other policy statements.

During the first decade of this century, Congress enacted four statutes that established a blueprint for what was to follow: The Biologics Act of 1902, the Federal Food and Drugs Act of 1906, the Federal Meat Inspection Act of 1906 and the Insecticide Act of 1910.

Of these four statutes, by far the most remarkable was the Biologics Act of 1902. In 1901, a contaminated smallpox vaccine caused an outbreak of tetanus in Camden and a single lot of tetanus-infected diphtheria antitoxin resulted in the deaths of several children in St. Louis. The Medical Society of the District of Columbia petitioned Congress to enact legislation to protect the public. Initially, it was limited to the District of Columbia. In the end, however, Congress changed it to cover the entire country.

The 1902 Act required premarket approval by the federal government of a product license application and an establishment license application for every biological drug intended for human use. This was the first time in history that any government had ever required explicit government licensing or approval of a category of consumer products before a product within that category could lawfully be marketed. Previous statutes in Europe, the American colonies and the states had prohibited adulterated or misbranded products, but had provided governmental authorities only with the power to police the marketplace—to review already-marketed products and to bring legal action against any product found to violate the statutory requirements. The 1902 Act thus represented the first time that any government was given the administrative authority to prevent the marketing of a consumer product simply by disapproving a marketing application or by taking no action at all on the matter. The same premarket authority was enacted into law for animal biological drugs a decade later, in the Virus, Serum and Toxin Act of 1913.

The premarket approval requirements of the Biologics Act of 1902 stood alone for more than 50 years. When Congress enacted the Federal Food and Drugs Act of 1906 to regulate the rest of the United States drug supply, it failed to authorize any form of premarket testing or approval, or even the development of administrative regulatory standards. The Federal Meat Inspection Act of 1906 and the Insecticide Act of 1910 similarly relied entirely upon traditional police powers and imposed no regulatory requirements prior to marketing.

Beginning in 1933, Congress considered replacement of the Federal Food and Drugs Act of 1906 with a more modern regulatory statute. As initially drafted, the legislation that ultimately became the Federal Food, Drug and Cosmetic Act of 1938 similarly included only policing authority. In the fall of 1937, however, a drug that had been hurriedly marketed with an untested solvent killed more than a hundred people within a few days. The solvent was diethylene glycol, a potent poison. In response to this tragedy, Congress included in the 1938 Act a new provision to require premarket notification of all new drugs. Even then, Congress stopped short of premarket approval. Under the 1938 Act, if a company submitted a New Drug Application (NDA) for a product and the Food and Drug Administration (FDA) took no action within 60 days, the NDA was deemed to be effective and the drug could lawfully be marketed.

A decade later, following World War II, Congress replaced the Insecticide Act of 1910 with the Federal Insecticide, Fungicide and Rodenticide Act (FIFRA) of 1947. For the first time, every pesticide was required to be registered before it could lawfully be marketed. The 1947 Act contained no authority, however, for the United States Department of Agriculture (USDA) to deny registration based upon an administrative determination that a product was adulterated or misbranded.

The combined impact of these various federal regulatory statutes on the Nation's economic vitality, industry's ability to develop and market new products for the benefit of the public, and individual freedom of choice, was negligible. Entrepreneurs could continue to develop, test and market their products in accordance with the statutory standards set forth in this legislation, without undue interference. Consumers could obtain the benefits of research quickly in the marketplace, without delay caused by government regulation. Products developed in the United States could reach the market as quickly as products developed abroad.

The marketplace policing—that is, postmarket regulation—by FDA and USDA under these regulatory statutes was extremely successful in weeding out adulterated and misbranded products. By taking strong regulatory action, these two agencies removed thousands of unsafe and ineffective products from the market, thus benefiting both consumers and the legitimate industry. Free choice in the marketplace was enhanced by mandatory label information required to be provided to consumers for regulated products. The premarket approval requirements for human and animal biological products had little impact because their scope was extremely narrow.

Beginning in the 1950s, however, the situation changed dramatically. Congress enacted a series of statutes that required premarket approval of a large number of consumer products:

- Miller Pesticide Amendments of 1954, requiring premarket approval of pesticide residues in or on food.
- Food Additives Amendment of 1958, requiring premarket approval of food additives.
- Federal Environmental Pesticide Control Act of 1972, requiring premarket approval of pesticides.
- Color Additive Amendments of 1960, requiring premarket approval of color additives.
- Drug Amendments of 1962, requiring premarket approval of new drugs.
- Animal Drug Amendments of 1968, requiring premarket approval of new animal drugs and feed additives.
- Medical Device Amendments of 1976, requiring premarket notification for all medical devices and premarket approval for Class III medical devices.
- Toxic Substances Control Act of 1976, requiring premarket notification for chemical substances.
- Infant Formula Act of 1980, requiring premarket notification for infant formulas.
- Nutrition Labeling and Education Act of 1990, requiring premarket approval of nutrient descriptors and disease prevention claims for food.

Taken either individually or collectively, this legislation has resulted in a revolution in product regulation. From simple policing prior to the 1950s, the federal government has gradually undertaken the role of sole gatekeeper to the marketplace.

Whereas the federal government formerly had responsibility for evaluating products for potential regulatory action after they were marketed, federal agencies now stand in the position of the sole decisionmaker as to whether a product will reach the market at all. For a product subject to premarket approval, no manufacturer may lawfully distribute the product, and no member of the public has a right to obtain it, unless and until the relevant federal agency authorizes marketing.

Over the years, regulation under these statutes has uniformly, inexorably increased in scope, detail and impact. When a regulatory statute is first enacted, the initial implementation is generally narrow and limited to the specific requirements of the law, and the impact is therefore modest. As time goes on, however, each successive administrator adds new requirements. None is ever revoked. The agency comes to regard the statute as broad and comprehensive, not narrow and limited. Regulation takes on a life of its own and becomes ever more burdensome, costly and intrusive.

This change in the regulatory function of our federal agencies has had a profound impact not only on the agencies themselves, but also on the country as a whole and each of us in our daily lives.

Premarket approval severely limits individual freedom of choice. Citizens are simply precluded from obtaining products they wish to purchase, and have no recourse other than to wait government approval. Personal autonomy is subjugated to government regulatory control. American citizens have been forced to travel abroad to obtain drugs and treatments not available here because of a slower and more stringent regulatory system in the United States.

Premarket approval has had enormous economic consequences. Particularly for small businesses, the investment required to obtain premarket approval of new products can be prohibitive. As approval requirements and government review times have increased, the cost of premarket approval has escalated. The greater the investment that is required, the higher the price that must be charged to consumers for the product when it is ultimately approved. American companies have been forced to test and market their products abroad, and to move manufacturing facilities overseas, as a result of increasingly onerous regulatory requirements in the United States.

Premarket approval includes no mechanism for public accountability. Citizens who wish to obtain a product have no right to participate in the process and no access to judicial review of whatever action is taken by the federal government. Even the applicant is precluded from access to the courts until final action is taken on a product application. Because the premarket approval process is conducted in private, without public participation, and without the right to judicial intervention or any other form of public accountability, it has been subject to increasing criticism.

In this book, focusing on biotechnology, Henry Miller astutely analyzes the impact of these developments. He documents various examples of scientifically flawed or merely ill-advised administrative policies and practices applied by federal agencies and supra-national organizations. He also describes the enormous ripple effects of governmental misapplication of regulation.

Biotechnology undoubtedly represents our greatest hope, as a society, for curing diseases for which we presently have no therapy, for providing optimal nutrition to an ever-expanding population and for solving myriad other societal problems. Public policy relating to biotechnology is therefore of vital interest to every citizen. This book describes and dissects that public policy, identifying flaws and shortcomings. More important, it offers realistic and intelligent solutions to them. It deserves your attention and thoughtful evaluation.

Peter Barton Hutt
Partner, Covington & Burling
Lecturer on Law, Harvard Law School
Former Chief Counsel, Food and Drug Administration

PREFACE

WHY I WRITE

When I graduated from MIT in 1969, I piled my worldly goods into my Pontiac and headed for graduate school in California, with a copy of James Watson's newly published *The Double Helix* in my suitcase. I never dreamt what the next quarter century would teach me about all manner of things, the wonder of being a scientist and being able, as Stanford biochemist and Nobelist Arthur Kornberg so elegantly captured it, "to frame a question and arrive at an answer that opens a window to yet another question and to do this in the company of like-minded people with whom one can share the thrill of unanticipated and extended vistas."

I could not have imagined what I would learn about the uncertain and sometimes uneasy relationship between basic science and product development; about government's somewhat schizophrenic roles in simultaneously funding and regulating research and development; and about the vagaries of the process under which the government establishes policies that affect the robustness, and indeed, the very existence of R&D.

I had learned at MIT to "internalize the canons of science" before I had ever heard the phrase. I had learned to make choices on a rational basis, whether I was selecting controls for an experiment or making lifestyle choices. I discovered that while science cannot help us to make moral decisions or value judgments, it is of value for helping to determine causation, quantify risks and validate the assumptions that underlie the paradigms used for protecting society from those risks.

I found that science is not a part-time diversion, like hang gliding or enophilia, or a societal "special interest" that occasionally clamors for attention or largesse from politicians or the public. Rather, it is a *modus vivendi* that enables one to order physical and biological phenomena; not only to organize and integrate what we know, but equally important, to understand the dimensions of what we don't know.

This volume draws heavily on essays and articles I have written over the past 15 years. My articles have appeared in a wide spectrum of outlets: peer-reviewed medical and scientific journals, textbooks and national newspapers and magazines. Most often, my writing was not part of any grand plan; rather, I was propelled to the word-processor by what I perceived to be stupidity, cupidity or the victimization of science and technology and their beneficiaries.

During that period "the new biotechnology" came of age. As a result of what is variously referred to as "genetic engineering," "gene splicing," "the new biotechnology" or "recombinant DNA" (rDNA) techniques—in which genes are identified, isolated and moved from one organism to another—more than a thousand products have been approved for marketing (the vast majority by the U.S. Food and Drug Administration) and untold numbers tested. Hardly any American's life has been untouched by them. These products include two dozen therapeutics like human insulin for diabetics; growth factors used in bone marrow transplants and to enhance cancer chemotherapy; tPA (tissue plasminogen activator) for treating heart attacks; hundreds of diagnostics tests for infectious agents (including those which cause AIDS and hepatitis) and for the detection or quantitation of drugs and hormones; and enzymes used to make cheese. Moreover, tens of millions of Americans have consumed dairy products from cows treated with the genetically engineered (that is, rDNA-derived) version of the hormone bovine somatotropin, or bST.

But that's just the beginning. There are new plant varieties created with the techniques of the new biotechnology that have begun to appear in markets and that have better taste, longer shelf life and a variety of other desirable traits.

"Human gene therapy," which introduces therapeutic genes into a patient's cells—for example, bone marrow, liver, skin, white blood cells or the lining of the lungs—is being tested in hundreds of patients throughout the world. These experiments are attempting to correct a variety of genetic defects, including those that cause very high cholesterol ("familial hypercholesterolemia") and cystic fibrosis; and also to treat AIDS and a variety of cancers. "Biotech" has become part of the vernacular (if you doubt it, check recent headlines in the *Wall Street Journal*). "Biotech companies" are a fixture of the landscape—and the stock exchanges.

Government policies toward the new biotechnology during the last decade have progressed uncertainly. On the positive side, oversight by the U.S. National Institutes of Health (NIH) has been relatively enlightened, with progressively decreasing stringency of their guidelines for laboratory research, and the elimination of NIH approvals of field trials.

However, from the beginning, a basic tenet of the NIH approach has been to treat recombinant DNA techniques as though they were unique with respect to risk, among all means of performing genetic manipulation. We have paid the price for that invalid assumption ever since. In crafting and adopting a *technique*-based regulatory mechanism—one emulated throughout most of the world—the NIH invented the wheel that would be rolled adroitly by generations of regulators and activist demagogues.

In contrast, the FDA's approach to the new techniques—that they were merely extensions, or refinements, of earlier techniques for genetic manipulation—led the agency to oversee products made with the new biotechnology in the same way as similar products made in other ways.

The FDA's rationale has been validated by near-unanimous scientific consensus and a decade of success; innumerable safe and effective products have made their way to market. Recently, however, FDA has shown signs of a sea change in policy, in the form of special regulatory requirements for foods and animals feeds made from rDNA-manipulated plants (chapter 3).

Good science and good sense have not always been well served by oversight of the new biotechnology, and too often the bureaucracy-builders have instituted burdensome regulation or retained it long after its time.

Planned introductions (field trials or "deliberate releases"), in particular, frequently have been overregulated. Several European countries and Japan have been unnecessarily restrictive, and they have paid the price in diminished research and development activity. International organizations such as the United Nations see in biotechnology R&D a potential bonanza—but not in the hopeful sense of new products for preventing disease and enhancing food quality and quantity. Rather, international bureaucrats anticipate a potential gravy train—new empires and budgets for regulators. Too often, cynical bureaucratic possibilities, rather than human needs, drive their policies.

In the United States, an equivocal or inhospitable regulatory climate and a lack of public understanding have discouraged research and development in certain important areas. In particular, overregulation has made academic and industrial researchers leery of field trials of organisms genetically engineered with the newest, most sophisticated techniques—even in areas such as bioremediation and pest-control, which are in high demand and which were once highly touted. As a result, investor interest in sectors that require field trials has diminished.

As a bench scientist, government regulator, and lately as an academic student of public policy, I have been privileged to play a role in some of these developments as they evolved. I hope that a few of them had a better outcome than had I not been involved. I never sought consciously to trade the sanctity of the laboratory bench for controversies over public policy issues, but I seem to have been drawn inexorably toward them.

This volume, which is organized into five chapters, touches on many of the controversies that have engulfed the new biotechnology during the past decade and, in fact, concentrates on those aspects of the new technology that have been the most contentious: the lack of perspective by the media and the public on the antecedents of the new biotechnology, governmental overregulation and its unfortunate effects, the negative influence of the anti-biotechnology groups and the race among various nations for preeminence in the marketing of products.

At the root of most major issues regarding biotechnology oversight today is the conflict between two opposing views of the basis of risk analysis. This is the "product versus process" controversy, discussed in chapter 1.

Regulators and others on the wrong side of this issue have ignored scientific consensus and allowed—even encouraged—myths about biotechnology to be perpetuated. This raises the question, if reason and a search for truth are not the basis for crafting expensive, intrusive and pervasive regulatory policy, what is? The answer currently seems to be politics, ideology and self-interest, as discussed in chapter 2.

The state of regulation within the agencies entrusted with primary oversight of biotechnology is discussed in depth in chapter 3, along with the complex systemic problems that inhibit enlightened rulemaking and effective performance. The United States is not alone is dealing less than admirably with the controversies of biotechnology. The effects of overregulation, while severe for industrialized nations, are more dire for developing nations: often lacking governmental oversight even of products or processes known to be hazardous, they can ill afford to expend resources on negligible- or low-risk activities.

International and supranational organizations have become part of the problem, not the solution. Far from ameliorating the social and technological problems of countries within the industrialized and developing world, the major treaties and regulatory codes being developed under the auspices of United Nations agencies will exacerbate them (chapter 4).

But the etiologies of deficient regulatory policy extend beyond government. The academic community has often been silent, while industry titans have colluded with regulators (excessive regulatory requirements can create market entry barriers to competitors). The most vociferous—and vicious—forces to express opinions on the subject have been anti-technology special interests.

Routes to a more rational basis for policy decision making do exist, however; these are discussed in chapter 5. There is no magic bullet. Improvement must be evolutionary, rather than revolutionary, and the substrate must be gradually enriched so that rational ideas will grow. We must have more involvement by scientific experts in public policy formulation; more aggressive leadership of the scientific community's institutions (journals, academies and professional associations); and greater understanding and interest on the part of public opinion leaders of the rudiments of science, technology and risk.

Sir Isaac Newton's statement, "If I have attained any heights, it is because I have stood on the shoulders of giants" may be a cliche, but it is certainly true for me. I wish to thank those giants, many of whom—Elkan Blout, Susanne Huttner, Donald Kennedy, Phil Leder, Frank Young—walk the earth still. Others, like Bernie Davis and Leonard Garren, are gone. One individual, John Cohrssen, who has served with distinction at the Council on Environmental Quality, the Office of the Vice President in the Bush administration and as a senior congressional advisor, deserves special mention. Courageous, competent, careful and always guided by the public interest, Cohrssen's influence is manifest again and again in this volume although his contributions are only occasionally singled out.

Finally, to any readers who may be embarking on careers in science or science policy, I offer some advice. British novelist Anthony Trollope said it is important for a young person entering life to decide whether he or she will make hats or shoes, but that is not half as important as the decision whether to make *good* or *bad* hats or shoes. My mentors have left me no choice in this, and I thank them.

ACKNOWLEDGMENTS

I thank Paula Duggan King for assistance in preparing and revising the manuscript. Dr. Susanne L. Huttner provided invaluable advice and stimulating discussions. Their expertise and forbearance made this book possible.

CHAPTER 1

Biotechnology's Seminal Issue: The "Product vs. Process" Controversy

ESSAY 1
BIOTECHNOLOGY POLICY'S SEMINAL ISSUE

Headlines in scientific journals and popular media trumpeted the arrival of the "biological revolution" and the transformation of research, industries and consumer products in ways never thought possible. Bacteria as factories, "genetically engineered" organisms, food products with characteristics unachievable through traditional breeding. Private investment in biotechnology companies soared and stock markets couldn't get enough of their new-found darling.

In the 1970s it all seemed too good to be true—and a handful of Cassandras in the environmental movement said it wasn't. They predicted imminent catastrophes and called for unprecedented governmental intervention into the testing and commercialization of the new biotechnology products. They urged restrictions even on early stage research in academic laboratories and called for case by case, every case review of experiments in the field. They described environmental and health risks of the magnitude associated with pandemics and nuclear meltdowns. Biotechnology, whether in research or in production, was heralded by both sides as something dramatically *different*, with unmeasured potential.

It wasn't long before biotechnology encountered the world of federal policy makers. As early as 1974, scientists convened an historic conference at Asilomar, California, that called on the National Institutes of Health (NIH)—the federal agency that promotes and funds biological research—to oversee the use of the new research methods.[1] The Asilomar conference focused on the potential risks of biotechnology's core technique, recombinant DNA (rDNA) technology, "gene-splicing." Discussions of risk management were as much influenced by a legitimate scientific debate about the likelihood of biosafety problems as by the brouhaha the meeting provoked in the national media. Frankenstein's monster and Andromeda Strain scenarios enlivened the news. Some of the assembled scientists believed that the public needed reassuring. They called on the NIH to develop a formal "biosafety" system to provide guidance to researchers and to ensure that laboratories employing the new techniques were properly equipped to prevent inadvertent release of rDNA-modified (genetically engineered) microorganisms.

When NIH established the Recombinant DNA Advisory Committee (RAC) and implemented the *Guidelines for Research with Recombinant DNA Modified Organisms*, the agency sent a powerful message that the scientific community was taking the popular risk scenarios seriously, a message that has affected biotechnology regulation worldwide. In an effort to reassure the public, the Guidelines were drawn in an overly cautious way. They promoted what has proved to be an idiosyncratic and largely invalid set of assumptions that overestimated the nature of risks associated with molecular biology techniques and rDNA-modified organisms. These assumptions have fueled a heated policy debate that continues unabated, today.

The seminal question about biotechnology regulation or, said another way, about government's approach to the risks of biotechnology, is: Should regulation focus on the genetic technique or on the characteristics of the resulting product? Is the use of rDNA techniques risky? Does the use of rDNA techniques affect product safety in a systematic way that should trigger a special oversight regime for the product? In order to protect the public safety, should government regulate rDNA research, or the end product, or both? Should the federal government create a new regulatory agency for all biotechnology activities, the philosophical equivalent of the airline industry's Federal Aviation Administration? Alternatively, should biotechnology products be treated like other similar products; that is, genetically engineered drugs regulated by the Food and Drug Administration (FDA), genetically engineered pest control agents regulated by the Environmental Protection Agency (EPA) and so forth?

Fast forward to the next decade. I was the FDA's primary reviewer for the first therapeutic drug produced by molecular genetic techniques when it came to the agency for market approval. The drug was human insulin, and the year was 1982.

From the outset, these genetic engineering techniques and the products made with them were referred to at the FDA as "the new biotechnology." This distinguishes them from older techniques such as mutagenesis selective breeding and chemical isolation of therapeutic molecules (like insulin) from human, animal or plant tissues. It was not intended to imply a fundamentally new or discrete category of products or to suggest that there were unique or inherent risks associated with the techniques used to produce them.

When the dossier for marketing approval for human insulin was submitted, we reviewed it according to what has proved to be an historically significant tenet: rDNA techniques are an extension and refinement of long-used and familiar methods for modifying organisms for various products and purposes. We were aware that these techniques enabled scientists to move genes between organisms at will. We knew, too, that they provided a mechanism for crafting genetic changes more precisely and obtaining more predictable outcomes. It was a time of great anticipation.

Based on a thorough review of data submitted by the manufacturer (Eli Lilly and Company), gained from years of preclinical testing in animals and clinical trials in diabetics, FDA granted marketing approval for human insulin. It took only five months at a time when the agency's average for new drugs was 30.5 months. This sent a message that was trumpeted on the *New York Times'* front page to biotechnology firms and their investors: biotechnology products would compete on a level playing field. Since 1982, untold numbers of new products have been tested and thousands have been approved for marketing, most of which are diagnostic tests (Table 1.1 lists the therapeutic drugs and vaccines approved for human use). Literally tens of millions of lives worldwide have been protected, enriched and even lengthened. Biotechnology applications have advanced into other fields, including agriculture and environmental cleanup. Hardly a week passes without press coverage of a significant biotechnology development in medicine or agriculture—the identification of genes involved in obesity, cystic fibrosis and cancer, or in fruit ripening, biological pest control and drought-resistance.

Used in the vernacular, "biotech" has become a catchword for an emerging entrepreneurial business sector and a wide spectrum of medical and agricultural products. The United States boasts world leadership in commercial biotechnology, with more than 1300 companies and $13 billion in sales revenues in 1995. Biotechnology is becoming commonplace in our daily lives, whether or not we are aware of it. High school students perform gene-splicing experiments in classrooms across the nation. Millions of American diabetics inject themselves daily with rDNA-derived human insulin, and hundreds of thousands of heart attack survivors can recount their experiences with the clot-dissolving drug tPA. Consumers are buying genetically engineered tomatoes in the supermarket, and a genetically engineered version of the protein, chymosin, is used to manufacture more than 60% of the cheese made in the United States.

These are some of the bright spots in a complex picture marked both by remarkable promise and shadows of controversy. The fate of new spinoff industries has turned on governmental decisions concerning the regulation of biotechnology research and products (Chapters 2 and 3). FDA's decision early on to treat new biotechnology products the same as other drugs and devices catalyzed frenzied investment activity in biopharmaceutical firms. Commercialization of agricultural and environmental products, by contrast, fared less well because other agencies decided to create new regulatory requirements and procedures for the new biotechnology. Why such different outcomes?

Early on, the aura of something approaching science fiction surrounded the new biotechnology. For an American public known for a high degree of risk aversion, biotechnology stirred both excitement and deeply rooted suspicions of the "new." By the mid-1980s, even as advances in biotechnology were becoming synonymous with the best in medical progress, rDNA techniques and biotechnology products were proving to be fodder for mythmakers and apocalyptics. The mythology quickly took root in agricultural

Table 1.1. Approved biotechnology drugs and vaccines

PRODUCT NAME	COMPANY	INDICATION (DATE OF U.S. APPROVAL)
Novolin® R regular, human insulin injection (recombinant DNA origin)	Novo Nordisk Pharmaceuticals (Princeton, NJ)	insulin dependent diabetes mellitus (July 1991)
Nutropin® somatropin for injection	Genentech* (S. San Francisco, CA)	growth failure in children due to chronic renal insufficiency, growth hormone inadequacy in children (March 1994)
Nutropin AQ™ somatropin (liquid)	Genentech* (S. San Francisco, CA)	growth failure in children due to chronic renal insufficiency, growth hormone inadequacy in children (December 1995)
OncoScint® CR/OV satumomab pendetide	CYTOGEN (Princeton, NJ)	detection, staging and follow-up of colorectal and ovarian cancers (December 1992)
ORTHOCLONE OKT®3 muromonab-CD3	Ortho Biotech* (Raritan, NJ)	reversal of acute kidney transplant rejection (June 1986); reversal of heart and liver transplant rejection (June 1993)
Proleukin® aldesleukin (interleukin-2)	Chiron (Emeryville, CA)	renal cell carcinoma (May 1992)
Protropin® somatrem for injection	Genentech* (S. San Francisco, CA)	human growth hormone deficiency in children (October 1985)
Pulmozyme DNase (dornase alpha)	Genentech* (S. San Francisco, CA)	cystic fibrosis (December 1993)
RECOMBI-NATE™ antihemophilic factor recombinant (rAHF)	Baxter Healthcare/ Hyland Division (Glendale, CA) Genetics Institute** (Cambridge, MA)	hemophilia A (December 1992)
RECOMBIVAX HB® hepatitis B vaccine (recombinant), MSD	Merck* (Whitehouse Station, NJ)	hepatitis B prevention (July 1986)
ReoPro® abciximab	Centocor (Malvern, PA) Eli Lilly* (Indianapolis, IN)	anti-platelet prevention of blood clots in the setting of high-risk PTCA (December 1994)
Roferon®-A interferon alfa-2a, recombinant	Hoffmann-La Roche* (Nutley, NJ)	hairy cell leukemia (June 1986); AIDS-related Kaposi's sarcoma (November 1988); chronic myelogenous leukemia (November 1995)

Table 1.1. Approved biotechnology drugs and vaccines (continued)

PRODUCT NAME	COMPANY	INDICATION (DATE OF U.S. APPROVAL)
Humatrope® somatropin (rDNA origin) for injection	Eli Lilly* (Indianapolis, IN)	human growth hormone deficiency in children (March 1987)
Humulin® human insulin (recombinant DNA origin)	Eli Lilly* (Indianapolis, IN)	diabetes (October 1982)
Intron A® interferon alfa-2b (recombinant)	Schering-Plough* (Madison, NJ)	hairy cell leukemia (June 1986); genital warts (June 1988); AIDS-related Kaposi's sarcoma (November 1988); hepatitis C (February 1991); hepatitis B (July 1992); malignant melanoma (December 1995)
KoGENate® antihemophiliac factor (recombinant)	Bayer Corporation, Pharmaceutical Division* (West Haven, CT)	treatment of hemophilia A (February 1993)
Leukine™ sargramostim (yeast-derived GM-CSF)	Immunex (Seattle, WA)	autologous bone marrow transplantation (March 1991); neutropenia resulting from chemotherapy in acute myelogenous leukemia (September 1995); allogenic bone marrow transplantation (November 1995); peripheral blood progenitor cell mobilization and transplantation (December 1995)
NEUPOGEN® Filgrastim (rG-CSF)	Amgen* (Thousand Oaks, CA)	chemotherapy-induced neutropenia (February 1991); autologous or allogeneic bone marrow transplantation (June 1994); chronic severe neutropenia (December 1994); support peripheral blood progenitor cell (PBPC) transplantation (December 1995)
Norditropin® somatropin (rDNA origin) for injection	Novo Nordisk Pharmaceuticals (Princeton, NJ)	treatment of growth failure in children due to inadequate growth hormone secretion (May 1995)
Novolin® 70/30 70% NPH human insulin isophane suspension & 30% regular, human insulin injection (recombinant DNA origin)	Novo Nordisk Pharmaceuticals (Princeton, NJ)	insulin dependent diabetes mellitus (July 1991)
Novolin® L Lente®, human insulin zinc suspension (recombinant DNA origin)	Novo Nordisk Pharmaceuticals (Princeton, NJ)	insulin dependent diabetes mellitus (July 1991)
Novolin® N NPH, human insulin isophane suspension (recombinant DNA origin)	Novo Nordisk Pharmaceuticals (Princeton, NJ)	insulin dependent diabetes mellitus (July 1991)

Table 1.1. Approved biotechnology drugs and vaccines (continued)

PRODUCT NAME	COMPANY	INDICATION (DATE OF U.S. APPROVAL)
Actimmune® interferon gamma-1b	Genentech* (S. San Francisco, CA)	management of chronic granulomatous disease (December 1990)
Activase® alteplase, recombinant	Genentech* (S. San Francisco, CA)	acute myocardial infarction (November 1987); acute massive pulmonary embolism (June 1990)
Activase® alteplase, recombinant (accelerated infusion)	Genentech* (S. San Francisco, CA)	acute myocardial infarction (April 1995)
Alferon® N interferon alfa-n3 (injection)	Interferon Sciences (New Brunswick, NJ)	genital warts (October 1989)
Avonex™ recombinant beta interferon 1a	Biogen** (Cambridge, MA)	relapsing multiple sclerosis (May 1995)
Betaseron® recombinant interferon beta-1b	Berlex Laboratories (Wayne, NJ) Chiron (Emeryville, CA)	relapsing, remitting multiple sclerosis (July 1993)
BioTropin human growth hormone	Bio-Technology General (Iselin, NJ)	human growth deficiency in children (May 1995)
Cerezyme™ imiglucerase for injection (recombinant glucocerebrosidase)	Genzyme (Cambridge, MA)	treatment of Gaucher's disease (May 1994)
Engerix-B® hepatitis B vaccine (recombinant)	SmithKline Beecham* (Philadelphia, PA)	hepatitis B (September 1989)
EPOGEN® Epoetin alfa (rEPO)	Amgen* (Thousand Oaks, CA)	treatment of anemia associated with chronic renal failure, including patients on dialysis and not on dialysis, and anemia in Retrovir®-treated HIV-infected patients (June 1989); treatment of anemia caused by chemotherapy in patients with non-myeloid malignancies (April 1993)
PROCRIT® Epoetin alfa (rEPO)	Ortho Biotech* (Raritan, NJ)	treatment of anemia associated with chronic renal failure, including patients on dialysis and not on dialysis, and anemia in Retrovir®-treated HIV-infected patients (December 1990); treatment of anemia caused by chemotherapy in patients with non-myeloid malignancies (April 1993)
[PROCRIT was approved for marketing under Amgen's epoetin alfa PLA. Amgen manufactures the product for Ortho Biotech.] Under an agreement between the two companies, Amgen licensed to Ortho Pharmaceutical the U.S. rights to epoetin alfa for indications for human use excluding dialysis and diagnostics.		
Genotropin™ somatropin (rDNA origin) for injection	Pharmacia & Upjohn* (Kalamazoo, MI)	short stature in children due to growth hormone deficiency (August 1995)

* PhRMA Member Company
** PhRMA Research Affiliate

Reprinted with permission from Pharmaceutical Research and Manufacturers Association.

and environmental applications of biotechnology. Demands were made for governmental protection against unseen and unmeasured (and often imaginary) risks.

In 1985, the Reagan Administration published a *Coordinated Framework for Regulation of Biotechnology* to clarify the regulations and procedures the major federal agencies had in place or were proposing for regulation of the new biotechnology.[2] This publication was a milestone, but it revealed a schism in federal policymaking among the agencies. Unlike FDA, EPA and the Department of Agriculture (USDA) found "gaps" in their existing policies and procedures that supposedly would leave the new biotechnology inadequately regulated.

EPA and USDA crafted new regulatory nets—that is, the scope of what would require their case by case review—on the assumption that the use of rDNA techniques, per se, creates *incremental* risk in new products. Five years earlier, FDA had judged that the vast majority of new biotechnology products were fundamentally similar to well-accepted and common products made with older, less sophisticated techniques. The FDA based this landmark decision partly on a consideration of first principles and also on extensive experience with microorganisms that had been genetically modified (using older techniques) for the production of therapeutic proteins, antibiotics and vaccines.

EPA and USDA rejected more than a century's experience with genetically modified plants used in crop breeding and decades' experience with genetically modified microorganisms used in biological pest control, mining, nitrogen fixation and bioremediation. Their actions implied that the new techniques and products were unfamiliar and warranted special governmental control.

Coming a decade after the NIH Guidelines, the not-so-well-coordinated Coordinated Framework added new fuel to the product-versus-process debate. While the debate narrowed to agricultural, food and environmental applications, the central question was the same: Are the molecular techniques of rDNA a straightforward extension of older genetic methods, or are they sufficiently different that they are likely to create novel and unfamiliar risks? If the former is true, then adequate protection could be achieved by EPA and USDA simply extending to the new biotechnology existing regulatory regimes that historically have overseen the use of older genetic techniques used to manipulate plants, animals and microorganisms. If the latter, then an entirely new risk analysis paradigm and new regulatory systems might be justified.

In the late 1980s, the United States National Academy of Sciences (NAS) and its research arm, the National Research Council (NRC), published a white paper[3] and special report,[4] respectively, that addressed risk analysis of rDNA-modified plants and microorganisms. In the argot of the report, effective risk management can be achieved for the vast majority of new biotechnology products in the same manner as for similar products of older genetic techniques. The report concluded that rDNA-modified plants and microbes typically present both a high degree of "familiarity" and a greater degree of certainty about genetic changes (see next essay).

The NAS and NRC findings resonated with FDA's earlier judgments. Genetically engineered human insulin (purified from *E. coli* which contained human DNA) is therapeutically similar to insulin extracted from cows or pigs, with the notable exceptions that it is purer and safer. The genetically engineered FlavrSavr tomato is virtually identical to other table variety tomatoes, except that it stays firm longer because the enzyme that naturally softens the fruit is better modulated.

Nonetheless, EPA and USDA proceeded with their plans for vastly greater case by case governmental review than for pre-rDNA genetically engineered products. The new regulations and proposals provided legally enforceable assurances to environmentalists and consumer activists, but at a high cost. The regulatory strictures caused companies and academic researchers to perform thousands of field experiments that measured or at least took into account the supposedly "unique" risks of rDNA-modified plants and microorganisms. Regulatory compliance directly and indirectly inflated the costs of performing routine field validation experiments. For the emerging commercial biotechnology sector, these policies raise four concerns, all of which can be considered "opportunity" costs:

1. By raising the cost of biotechnology R&D, the new EPA and USDA regulatory schemes drain capital resources and slow the pace of research. This stalls innovation, which, in turn, delays or blocks altogether the entry of new products into the marketplace. Viewed from a competitive business perspective, it can sustain reliance on less efficient, less precise, less predictable and sometimes more hazardous alternative technologies and products.

2. Viewed from the perspective of the financial community, the higher operating costs and extended development times associated with using rDNA techniques raise investment risk and exacerbate concerns about long term prospects for company success. This leads to depressed valuation of start-up firms and subsequent financial offerings. Less capital and higher "burn rates" jeopardize smaller firms' ability to achieve R&D milestones. Agricultural and environmental biotechnology companies find capital streams drying up or completely unavailable, in contrast to their counterparts in the biopharmaceutical sector (chapter 2).
3. The new regulations disproportionately affect the academic research community, an important historical source of fundamental scientific knowledge, improved germplasm, and the highly skilled workforce on which biotechnology companies depend. The costs of performing and submitting required environmental assessments divert funds from already strapped research budgets. Many scientists choose not to pursue early field research with genetically engineered plants and microorganisms because their resources are spent more cost-effectively on less onerously regulated laboratory research. Many consider the kinds of field "experiments" permitted to be of limited scientific or heuristic value and to provide a poor training environment for graduate students and postdoctoral trainees. (And certainly, the paperwork involved in obtaining government permits is an unwelcome diversion.) Therefore, researchers forego field validation research that is essential in order to develop plants and microorganisms for use in agriculture or environmental bioremediation. Without this needed proof of concept, biotechnology innovations remain on the bench. They are not transferred to the commercial R&D pipeline and, therefore, remain out of reach of end-users and consumers.
4. Viewed from the perspective of vulnerability to legal challenges, delays and disruptions, EPA and USDA regulations provide legal avenues by which an individual or group ideologically opposed to biotechnology generally

> or to a particular application can intervene, relatively easily, in the process of research and commercialization. Putting it another way, an unnecessarily wide regulatory net exposes scientists and companies to publicity (because agencies are most often required to publish their decisions on individual products). The researchers, manufacturers and regulatory agencies, alike, then become vulnerable to petitions, lawsuits, demonstrations and boycotts, on a product by product or experiment by experiment basis.

Taken together, USDA and EPA regulatory schemes have tilted the playing field against researchers and companies that use the techniques of the new biotechnology. In doing so, they have defied the primary goal of the United States government's own Coordinated Framework, which was to limit potential product risks while *encouraging innovation and economic development.*[5] During public meetings, USDA and EPA staff have defended themselves as purveyors of public confidence for new biotechnology products. At a 1991 conference in Sacramento, California, Charles Hess, then USDA Assistant Secretary for Science and Education, said that if the public *believes* a risk is real, then the government has a responsibility to regulate it. Such a philosophy only serves to enhance antibiotechnology (and more general antitechnology) mythology.

ESSAY 2
MYTHS ABOUT BIOTECHNOLOGY

There is a place in society for myths. Like Parson Weems' parable about George Washington chopping down the cherry tree ("I cannot tell a lie..."), they illustrate simple moral lessons. Arguably, learning about fictional heroes from moral fables is, itself, empowering. When misinformation is cloaked in a pretense of reality, however, myths can mislead, misinform and undermine social goals. Biotechnology has been shrouded by two basic myths: that it is a discrete technology characterizing a uniform set of activities and products, and that its effects are unpredictable, unfamiliar and likely to spawn dangerous organisms. These myths are born of concerns about technology generally, and of limited understanding of genetics and biology in particular.

Myth One: Biotechnology is a Discrete and New Technology

Biotechnology is, in fact, not a discrete technology. It refers to a group of useful enabling technologies, including rDNA and hybridoma techniques, that have wide application in research and commerce. Over the past several decades, they have become so integrated into the practice of plant breeders and microbiologists and so commingled with conventional techniques as to make distinctions between old and new meaningless.

A useful working definition of biotechnology used by several United States government agencies is "the application of biological systems and organisms to technical and industrial processes."[1] It encompasses advances in biology, genetics and biochemistry that are applied to processes as different as drug development, fish farming, forestry, crop development, mining and oil spill cleanup. It refers to discovery and developmental research, as well as to manufacturing. The universe of biotechnology is so diverse as to defy a conceptual construct that cohesively binds the components together. The absence of systematic, uniform characteristics further defies efforts to prescribe the kinds of overarching regulatory systems that have been applied to specific industrial sectors, such as coal mining or commercial airplane traffic.

Biotechnology spans no less than three major federal regulatory agencies, whose mandates have been variously defined by the Congress according to: product end use (e.g., FDA's jurisdiction under the Food, Drug and Cosmetic Act, over drugs, devices, foods and cosmetics); risk (e.g., USDA's jurisdiction under the Federal Plant Pest Act, over "plant pests"); or "novelty" (e.g., EPA's jurisdiction under the Toxic Substances Control Act, over new substances, or mixtures thereof, used for a wide variety of applications, including bioremediation and mining). As the regulatory mandate varies, so does the nature of the agencies' risk assessment and management schemes. Consider, for example, a proteolytic enzyme effective both as a drain cleaner and in dissolving blood clots in arteries. EPA's review of the drain cleaner would be very different from the exhaustive animal testing and clinical trials FDA would require for the blood clotting drug.[2]

The diversity of products and applications, coupled with the agencies' divergent historical approaches to regulation, argues against the usefulness of a unified legislative approach to biotechnology regulation. Such an approach would force artificial and arbitrary groupings of products, such as "biotechnology," "genetically engineered" or "genetically manipulated organisms." It would create redundancies and overlap in agency jurisdictions. It would likely distance regulatory decision-making from decades of regulatory experience at the existing agencies. It would create inconsistencies between regulatory requirements for products with similar or even virtually identical properties that differ only in the techniques used to create them. These factors, taken together, would seriously undermine effective risk analysis.

The other issue, the "newness" of biotechnology, is best considered from an historical perspective, as discussed briefly in the previous essay. Biotechnology has been in practice for millennia. As early as 6000 B.C., Sumerians and Babylonians used yeast to brew beer. An Egyptian tomb contains a painting of people preparing and fermenting grain and storing the brew.[3] In domesticating microorganisms and enhancing their desired physical characteristics (phenotypes), these ancient "biotechnologists" unknowingly but nonetheless systematically pursued genetic modifications: genetic changes (that is, new genotypes) evolved, unseen, as new strains or varieties were selected.

In modern times, the science of genetics has been applied to produce many valuable variants of yeast and bacteria. For example, the capacity of the mold *Penicillium chrysogenum* to produce penicillin has been increased more than 100-fold during the past several decades.[4] One of my favorite microorganisms, *Lactobacillus san francisco,* is a bacterium used to make sourdough bread. Other microorganisms have been selected and genetically enhanced in their ability to produce foods, beverages, industrial detergents, antibiotics, organic solvents, vitamins, amino acids, polysaccharides, steroids and vaccines. A decade ago, these (pre-rDNA) biotechnology products, together, had a value in excess of $100 billion annually.[5] Recombinant DNA techniques have provided both an important new set of tools for microbiologists and access to a broader range of markets. They enable researchers to precisely identify, characterize, enhance and transfer individual genes involved in biosynthesis of target products. They also enable biopharmaceutical companies to use bacteria as biological factories for synthesizing therapeutic substances like human insulin, human growth hormone and erythropoietin. The long search for useful microorganisms begun by the ancients is now driven by a continuum of biotechnologies, which has culminated in various manifestations of rDNA techniques.

Many people are surprised to learn that one of the ubiquitous uses of (pre-rDNA) genetically modified organisms is the vaccination of human and animal populations with live, attenuated viruses. Live viruses modified by various techniques to reduce or limit their disease-causing capabilities have been used for decades in vaccines against mumps, measles, rubella, smallpox, poliomyelitis and yellow fever. Inoculation of a live viral vaccine entails the controlled "infection" of the recipient in an effort to stimulate the immune system's protection against subsequent uncontrolled infections. Even attenuated live viruses present the possibility of further transmission of the virus in the community. In theory, there is a risk of serial propagation in the environment through human or animal hosts in the community. In spite of the continual presence of vaccine viruses around the world, there is no evidence that a vaccine virus has become established and propagated in the environment. Studies in the U.S. and the United Kingdom have demonstrated that vaccine strains of poliovirus are present in sewage as a result of the continuing administration and excretion of live virus vaccine, rather than its serial propagation.[6] Viral vaccines produced with these older genetic techniques have been used remarkably effectively throughout the world and have completely eradicated the dread smallpox virus.

Practical applications of (pre-rDNA) biotechnology, of course, extend much further. In agriculture, they include a variety of organisms used in pest control (including many often considered to be pests, themselves, in other settings). Insect release was used successfully to control troublesome weeds in Hawaii in the early 20th century and St. Johnswort ("Klamath weed") in California in the 1940s and 1950s. More recently, an introduced rust pathogen has been used to control rush skeletonweed in Australia. Hundreds of successful planned introductions of natural predators, such as the Australian Vedalia beetle in California in 1988, have made research into biological control of weeds, nematodes, insects and diseases a priority at USDA and the U.S. Department of Interior.[7] Currently, several dozen microbial biocontrol agents, including bacteria, viruses and fungi—in hundreds of formulations—have been approved and registered with the EPA. They control a broad spectrum of important plant pests and provide effective alternative strategies to increasingly unavailable chemical pesticides.

Biological agents are also used as growth promoters for plants. Preparations containing the bacterium *Rhizobium* have been sold in the U.S. since the late 19th century. These bacteria stimulate the development of plant root structures that, in turn, enable the bacteria to convert nitrogen into compounds needed by the plant. The bacteria "fix" atmospheric nitrogen, converting it into nitrogen-containing ions that are essential plant nutrients, and thereby enhance the growth of leguminous plants such as soybeans, alfalfa and beans. The use of nitrogen-fixing microorganisms (or, potentially, the introduction of the genes responsible for nitrogen-fixation directly into the plant) decreases the need for chemical fertilizers.

Then there is the agricultural "green revolution" which has dramatically increased human longevity and improved the quality of life in developing countries. The green revolution might be viewed as the culmination of a long quest begun by ancient agriculturists who attempted to cultivate and domesticate wild plants. With the rediscovery in 1900 of Gregor Mendel's concepts of inheritance, plant breeders ushered in the era of scientific application of genetic principles to crop improvement. Twentieth century plant breeding, even before the advent of rDNA methods, sought ways to take advantage of useful genes and has found a progressively wider range of plant species and genera on which to draw. Breeders first achieved interspecies hybridization, transferring genes between different, but related species. Eventually, plant geneticists found ways to perform even wider crosses between members of different genera. The offspring of such crosses normally are not viable because the resulting embryo has an abnormal endosperm. With the development of tissue culture techniques, however, the hybrid embryo could be provided conditions similar to those supplied in early development by the normal endosperm and maternal tissues. Crops resulting from such "wide crosses" are commonly grown and marketed in the U.S. and elsewhere. They include familiar and widely used varieties of tomato, potato, corn, oat, sugarbeet, bread and durum wheat, rice and pumpkin.

Recombinant DNA techniques have refined these methods by enabling plant breeders to identify and transfer *single* genes encoding specific traits of interest. They can now readily move selected and well characterized genetic material from any source in nature, greatly increasing the diversity of useful genes and germplasm available for crop improvement. In addition, safety assessment of plants is enhanced by the greater precision and predictability of rDNA techniques.

Myth Two: Novel and Dangerous Organisms will be Created

Does the introduction of a single gene or several genes, judged against the background of tens or hundreds of thousands of the host organism's own genes, create a "novel" organism? Could the introduction of a single gene from a known pathogen convert an otherwise benign organism into a pathogen?

How novel is a corn plant with a newly-inserted gene for a *Bacillus thuringiensis* endotoxin which confers endogenous protection against European corn borer? It is, after all, still a corn plant. How novel is the laboratory bacterium *Escherichia coli (E. coli)* K-12 with a newly-inserted gene for human alpha interferon? It only varies from the unmodified strain in one gene. Does a tomato plant become pathogenic (or give rise to novel pathogenic recombinants) after insertion of a small amount of well characterized DNA from *Agrobacterium tumefaciens*, a known plant pathogen? These questions have been widely debated and answered, as reflected by the publications of the NAS, NRC and other groups described in the following essay.

Consider, first, whether genetic recombination, itself, is of concern. It has already been established that people have for millennia engaged in genetic recombination in the production of domesticated microbes, plants and animals. But the impact and importance of these genetic changes pale in comparison with what occurs in nature. Innumerable recombinations between related and unrelated organisms have occurred by several mechanisms.[8] Sexual reproduction randomly combines genes from two parents in the offspring, which then has a unique set of genes to pass along to the next generation. In the gut, decomposing corpses, and infected wounds, bacteria take up naked mammalian DNA, albeit inefficiently, when they encounter disintegrating cells. Over the past million years and longer, mammalian-bacterial genetic hybrids have appeared, been tested by competition within bacterial populations and by environmental stresses, and have been conserved or discarded by natural selection. Genetic hybridization also has been rampant among fungi, viruses and plants.

Similar to the wide crosses in plants described above, certain kinds of gene transfers thought until recently to be impossible in nature because of phylogenetic distances are now known to occur. Brisson-Noel et al have demonstrated that a gene (or genes) for erythromycin resistance was transferred between the gram-negative bacterium *Campylobacter* and unrelated gram-positive bacteria.[9] In recent laboratory experiments, it was demonstrated that gene transfer can occur between *E. coli* and streptomyces[10] or yeast,[11] and that the crown gall disease in plants results from a natural transfer of DNA from *Agrobacterium* to plant cells. In fact, knowledge of the nature of this mechanism led to the development of an effective means of transferring selected genes into plants using *Agrobacterium* DNA.[12]

Evolutionary studies provide data relevant to the issue of the "novelty" of molecular chimeras created by rDNA gene-splicing. Does the transfer of a moth gene into a squash affect the "squashness" or transfer "mothness" to the new hybrid? The sequencing of various genomes during the past decade has revealed that nature has been remarkably conservative about maintaining and using effective molecules as they evolved. Nearly identical DNA sequences and biochemical pathways are found in different species, across genera, and even across phylogenetic kingdoms. Scanning the sequence of the *E. coli* genome, for example, reveals gene sequences that are virtually identical to those in a variety of organisms, ranging from other bacteria to plants, insects, amphibians and humans.[13] With such broad conservation and "sharing" of genes in nature, debates over the proprietary nature of "human," "plant," and "bacterial" genes and over the "novelty" of a squash plant that contains a moth gene become moot.

Taken together, the evidence on genetic recombination and evolutionary conservation of genes make distinctions drawn between "natural" and "unnatural," or "familiar" and "novel," seem neither clear nor relevant.

The second issue, conversion of a nonpathogen into a pathogen through limited genetic recombination, is best considered within the context of the natural phenomena that underlie pathogenicity. This process is both complex and multifactorial. Pathogenicity is not a *trait* produced by a single gene. Rather, it requires the coordinated activity of a set of genes that affect essential properties.

A pathogen must possess three general characteristics, each of which involves multiple genes. First, it must survive and multiply in or upon host tissues. The oxygen tension and pH must be satisfactory, the temperature (and for plant pathogens, the tissue water potential) must be suitable, and the nutritional milieu must be favorable. The pathogen must be able to adhere to specific surfaces on or in the host and thrive on nutrients available in the host environment. Second, the pathogen must be able to resist or avoid the host's defense mechanisms for the period of time necessary to produce numbers of offspring sufficient to cause disease. For human and animal pathogens, this includes resistance to enzymes, antibodies and phagocytic cells in the host. Third, the pathogen must be able to survive outside the host and must be disseminated to new host organisms.

The organism must be meticulously adapted to this pathogenic lifestyle. A single gene, even one encoding a potent toxin, will not convert a harmless bacterium into a pathogen capable of causing epidemics or even localized disease. On the other hand, a mutation that interferes with a gene essential to one of the three characteristics of a pathogen can *eliminate* pathogenicity. It is worth noting that severe pathogenicity is even more dependent upon favorable conditions and is, therefore, much rarer in nature than mild pathogenicity.

A similar situation pertains to weediness in plants. Analogous to the multifactorial nature of pathogenicity, weed biologists have discovered 13 characteristics that contribute to weediness; most serious crop weeds have 11 or 12 of these. Crop plants, in contrast, have only 5 or 6 of these 13 characteristics.[14] Because each of these characteristics is coded for by one or perhaps multiple genes, it is unlikely that a crop plant could suddenly acquire the genetic information necessary to transform it into a weed. A corollary is that the introduction into a crop plant of a single (or several) gene(s) unrelated to weediness is unlikely to confer weediness.

The probability of inadvertently creating an organism capable of producing a medical or agricultural catastrophe, therefore, is vanishingly small. There is no support for the notion promoted by some biotechnology critics that genetic recombination of the sort achieved with rDNA techniques will create pathogens from nonpathogenic organisms (or significant weeds from nonweedy organisms). Moreover, there is no scientific logic to be found in assertions that although the likelihood is small that a problem would be created through the use of rDNA techniques, the outcome should one occur will be cataclysmic. Such cataclysmic effects would depend upon the highly improbable creation of severe pathogenicity in a nonpathogenic organism by adding or changing one or a few genes. Just as in other applications of technology, such as scissors, automobiles and analgesic drugs, when mishaps occur they are usually of low impact rather than catastrophic.

The record of safety in using rDNA techniques in thousands of laboratories worldwide is illustrative. For more than 20 years, in spite of releases measuring on the order of 10^8 recombinant microorganisms per worker per day from standard Biosafety Level 1 (BL 1) laboratories,[15] not a single adverse reaction has been observed in humans, animals or the environment. It is nonetheless instructive to examine the scientific literature on real-life, worst-case scenarios for handling potentially harmful organisms: the effects on the community of laboratory-acquired infections in personnel exposed to highly pathogenic organisms. The NIH/Centers for Disease Control (NIH/CDC) handbook, "Biosafety in Microbiological and Biomedical Laboratories," summarizes the data:

> ...[L]aboratories working with infectious agents have not been shown to represent a threat to the community. For example, although 109 laboratory-associated infections were recorded at the Center for Disease Control in 1947-1973, no secondary cases were reported in family members or community contacts. The National Animal Disease Center has reported a similar experience, with no secondary cases occurring in laboratory and nonlaboratory contacts of 18 laboratory-associated cases, occurring 1960-1975. A secondary case of Marburg disease in the wife of a primary case was presumed to have been transmitted sexually 2 months after his dismissal from the hospital. Three secondary cases of smallpox were reported in two laboratory-associated outbreaks in England in 1973 and 1978. There were earlier reports of six cases of Q fever in employees of a commercial laundry which handled linens and uniforms from a

> laboratory where work with the agent was conducted, one case of Q fever in a visitor to a laboratory, and two cases of Q fever in household contacts of a rickettsiologist. *These cases are representative of the sporadic nature and infrequent association of community infections with laboratories working with infectious agents (emphasis added).*[16]

The NIH/CDC handbook describes the experience with mishaps using the most pathogenic and virulent organisms known, many of which have been studied for potential use in biological warfare. It is within this context that some biotechnology critics would place obviously innocuous domesticated plants, animals and microorganisms that have been precisely and minimally modified with rDNA techniques.

Fear of the Unknown

The myths surrounding newness, novelty and acquired pathogenesis are an expression of a fear of what is unknown about the new biotechnology—and new technologies, in general. Some have argued that with new biotechnology products, the unknowns far outweigh the knowns.[17] What matters is not what is unknown, per se, but what is unknown that is related to risk. The scientific method and prior experience applied in a logical fashion to risk analysis have provided a rational system through which one can make useful predictions about potential risks—and about the importance of what is unknown.

For example, with respect to mutations induced to create live poliovirus vaccines, it might be literally true that "the unknowns outweigh the knowns." Yet, the fact that the genetic changes achieved through mutagenesis are largely uncharacterized has not prevented our using the vaccines with monumental success over the past four decades. Likewise, for thousands of new plant varieties and other microorganisms of commercial interest, there is much that we do not know about them, but we have benefited from their use nonetheless. In each case, practitioners have systematically developed standard practices to ensure the quality of their products and the safety of their research and manufacturing practices.

Unless they are employed in research or industry, members of the public have little reason to know about these safeguards. It is a popular belief that new technology is dangerous. This view may be an expression of an atavistic fear of disturbing the natural order or of breaking primitive taboos. It is probably little affected by scientists' statistics-based that a new technology or product is safe. Few people understand the relationship between mathematical probability and risk. The media, always eager to find a story of technology-run-amok, are no help.

What is somewhat surprising is that technophobia thrives in the face of the believer's own direct experience with unequivocal successes like telephonic communication, vaccination, air travel, therapeutic drugs and diagnostics, microprocessors, and foods produced by genetically improved plants, animals and microbes. In earlier eras, techno-skeptics predicted electrocution from the first telephones, believed Jenner's early attempts at small pox vaccination would create monsters, and doubted the possibility of matching blood for transfusions. Today, they warn of "Andromeda Strains" and "Jurassic Parks" resulting from rDNA-modified organisms. Then and now, the believers have said that the costs would be too high.

No responsible person would deny that certain of the practices, processes or products of biotechnology are potentially hazardous under specific conditions. Workers purifying antibiotics have experienced allergic reactions. Laboratory workers have been inadvertently infected by pathogenic bacteria. Vaccines have occasionally elicited serious adverse reactions. The infrequency of these occurrences can be attributed to the application of the scientific method to assessing and managing risk. That method relies heavily upon both first principles and empirical evidence.

The scientific method is subverted, however, when inaccurate assumptions are used, when experiments are ill-conceived or poorly-performed. ("Garbage in, garbage out.") Consider the way in which the risks of rDNA modified organisms are commonly analogized to those associated with the introduction of exotic species—an invalid assumption.

There is clear evidence that introductions of exotic species such as English sparrows, gypsy moths and kudzu vines have had seriously detrimental ecological and economic consequences. In their new settings, these organisms were no longer constrained by controlling factors found in their native environments. Competing effectively in their new environments, they won the evolutionary lottery and thrived. This enhanced competitiveness does not depend upon one or a few genes, but rather on complex traits and genetic systems. Thus, exotic species do not present an appropriate model for judging the likely risks of introducing an rDNA-modified domesticated organism that has had one or a few genes modified. Domesticated organisms have been systematically modified to enhance commercially important traits that are commonly of little advantage in the native environment and often place the organism at a competitive disadvantage. Small, discrete genetic modifications of domesticated species simply cannot achieve the sort of adaptive advantage presented by the exotic species. A scheme for regulatory oversight based on experience with exotic species would unnecessarily burden development and utilization of modified domestic species.

What is the appropriate analogy for rDNA-modified organisms? It would combine knowledge of the unmodified domesticated parental organism, the trait(s) conveyed by the transferred gene(s), and the expected ways in which the modified organism is to be used (including familiarity with the environment into which it will be released).

It is seldom recognized that most residents of industrialized countries have had extensive experience with all manner of (pre-rDNA) biotechnology products—the result of continual contact with food, fiber and medical products derived from organisms that have been enhanced in some way. Applying this knowledge and experience, especially as they have been developed and archived at regulatory agencies and in the scientific literature, will ensure that new biotechnology products are appropriately evaluated for potential risks to consumers and the environment. Equally important, it will ensure that research and development are not unnecessarily burdened by the direct and indirect costs of regulation.

The goals of effective regulation are always coherence, logic and clarity, not leniency. Rationalized regulation will allow biotechnology products to compete on a level playing field with other, similar products. It will also provide consumers with a wider variety of product choices at reasonable prices.

ESSAY 3
BROAD SCIENTIFIC CONSENSUS

For two decades, international scientific organizations and professional groups have grappled with the correct assumptions on which to base regulation of the new biotechnology. To the pivotal question whether the use of the new techniques is sufficient to require a new regulatory paradigm, their answer has been virtually unanimous: *No.*

Without intending to belabor the point, it is worth describing the consensus in some detail, because of the striking congruence in the actual conclusions and recommendations.

The United States National Academy of Sciences (NAS) published in 1987 a white paper on the planned introduction of genetically modified organisms into the environment.[1] It has had wide-ranging impacts in the United States and internationally. Its most significant conclusions and recommendations include:

1. Recombinant DNA techniques constitute a powerful and safe new means for the modification of organisms;
2. Genetically modified organisms will contribute substantially to improved health care, agricultural efficiency and the amelioration of many pressing environmental problems that have resulted from the extensive reliance on chemicals in both agriculture and industry;
3. There is no evidence of the existence of unique hazards either in the use of rDNA techniques or in the movement of genes between unrelated organisms;
4. The risks associated with the introduction of rDNA-engineered organisms are the same in kind as those associated with the introduction of unmodified organisms and organisms modified by other methods and
5. The assessment of risks associated with introducing recombinant DNA organisms into the environment should be based on the nature of the organism and of the environment into which the organism is to be introduced, and independent of the method of engineering per se.

In a 1989 extension of this white paper, the United States National Research Council (NRC), the research arm of the NAS, concluded that "no conceptual distinction exists between genetic modification of plants and microorganisms by classical methods or by molecular techniques that modify DNA and transfer genes," whether in the laboratory, in the field or in large-scale environmental introductions.[2] The NRC report supported this statement with extensive observations of past experience with plant breeding, and of introduction of genetically modified plants and microorganisms:

- "The committees [of experts commissioned by the NRC] were guided by the conclusion (NAS, 1987) that the *product* of genetic modification and selection should be the primary focus for making decisions about the environmental introduction of a plant or microorganism and not the *process* by which the products were obtained" (p. 14).
- "Information about the process used to produce a genetically modified organism is important in understanding the characteristics of the product. However, the nature of the process is not a useful criterion for determining whether the product requires less or more oversight" (p. 14).
- "The same physical and biological laws govern the response of organisms modified by modern molecular and cellular methods and those produced by classical methods" (p. 15).
- "Recombinant DNA methodology makes it possible to introduce pieces of DNA, consisting of either single or multiple genes, that can be defined in function and even in nucleotide sequence. With classical techniques of gene transfer, a variable number of genes can be transferred, the number depending on the mechanism of transfer; but predicting the precise number or the traits that have been transferred is difficult, and we cannot always predict the phenotypic expression that will result. With organisms modified by molecular methods, we are in a better, if not perfect, position to predict the phenotypic expression" (p. 13).
- "With classical methods of mutagenesis, chemical mutagens such as alkylating agents modify DNA in essentially random ways; it is not possible to direct a mutation to specific genes, much less to specific sites within a gene. Indeed, one common alkylating agent alters a number of different genes simultaneously. These mutations can go unnoticed unless they produce phenotypic changes that make them detectable in their environments. Many mutations go undetected until the

organisms are grown under conditions that support expression of the mutation" (p. 14).

- "Crops modified by molecular and cellular methods should pose risks no different from those modified by classical genetic methods for similar traits. As the molecular methods are more specific, users of these methods will be more certain about the traits they introduce into the plants" (p. 3).
- "The types of modifications that have been seen or anticipated with molecular techniques are similar to those that have been produced with classical techniques. No new or inherently different hazards are associated with the molecular techniques. Therefore, any oversight of field tests should be based on the plant's phenotype and genotype and not on how it was produced" (p. 70).
- "Established confinement options are as applicable to field introductions of plants modified by molecular and cellular methods as they are for plants modified by classical genetic methods" (p. 69).

The NRC proposed that the evaluation of experimental field testing be based on three considerations: familiarity, i.e., the sum total of knowledge about the traits of the organism and the test environment; the ability to confine or control the spread of the organism, and the likelihood of harmful effects if the organism should escape control or confinement.

The same principles were emphasized in the comprehensive report by the United States National Biotechnology Policy Board (on which I served as a charter member), which was established by the Congress and comprised of representatives from the public and private sectors. The report concluded that:

> "[t]he risks associated with biotechnology are not unique, and tend to be associated with particular products and their applications, not with the production process or the technology per se. In fact biotechnology processes tend to reduce risks because they are more precise and predictable. The health and environmental risks of not pursuing biotechnology-based solutions to the nation's problems are likely to be greater than the risks of going forward."[3]

These findings resonate with the observations and recommendations of the United Kingdom's House of Lords Select Committee on Regulation of the UK Biotechnology Industry and Global Competitiveness,[4] which are discussed below. The Paris-based Organization for Economic Cooperation and Development (OECD), approaching the question from the perspective of the safety of new varieties of foods, came to similar conclusions in a 1993 report, "Concepts and Principles Underpinning Safety Evaluation of Foods Derived by Modern Biotechnology."[5] The OECD report described several concepts related to food safety that are consistent with the NRC and related findings:

- "In principle, food has been presumed to be safe unless a significant hazard was identified."
- "Modern biotechnology broadens the scope of the genetic changes that can be made in food organisms and broadens the scope of possible sources of foods. This does not inherently lead to foods that are less safe than those developed by conventional techniques."
- "Therefore, evaluation of foods and food components obtained from organisms developed by the application of the newer techniques does not necessitate a fundamental change in established principles, nor does it require a different standard of safety."
- "For foods and food components from organisms developed by the application of modern biotechnology, the most practical approach to the determination of safety is to consider whether they are *substantially equivalent* to analogous conventional food product(s), if such exist."

Various other national and international groups have repeatedly echoed or extended these conclusions; their observations are described briefly below:

- A joint statement from the International Council of Scientific Unions' (ICSU) Scientific Committee on Problems of the Environment (SCOPE) and the Committee on Genetic Experimentation (COGENE) (Bellagio, Italy 1987):

  "[t]he properties of the introduced organisms and its target environment are the key features in the assessment of risk. Such factors as the demographic characterization of the introduced organisms; genetic stability, including the potential for horizontal transfer or outcrossing with weedy species; and the fit of the species to the physical and biological environment...apply equally to both modified or unmodified organisms; and, in the case of modified organisms, they apply independently of the techniques used to achieve modification. That is, it is the organism itself, and not how it was constructed, that is important."[6]

- The report of a NATO Advanced Research Workshop (Rome 1987):

  "In principle, the outcomes associated with the introduction into the environment of organisms modified by rDNA techniques are likely to be the same in kind as those associated with introduction of organisms modified by other methods. Therefore, identification and assessment of the risk of possible adverse outcomes should be based on the nature of the organism and of the environment into which it is introduced, and not on the method (if any) of genetic modification."[7]
- The report of the UNIDO/WHO/UNEP Working Group on Biotechnology Safety (Paris 1987): "[t]he level of risk assessment selected for particular organisms should be based on the nature of the organism and the environment into which it is introduced."[8]

ESSAY 4
RISK-ASSESSMENT EXPERIMENTS

The conclusory statements by scientific academies and professional groups, based largely on theoretical principles, were not the only available "evidence" on which to base conclusions. In the late 1970s, when the regulation of recombinant DNA experimentation was at its most restrictive, it was decided that a bona fide risk assessment experiment to test one worrisome scenario would ameliorate concerns that had been expressed. This scenario was that cloning the DNA of an animal tumor virus into a bacterium that normally resides in the human gut could produce human cancer—

with the potential of an epidemic, should the bacterium escape from the laboratory. Consequently, elaborate experiments were carried out to test whether a disabled strain of recombinant *E. coli* containing the entire genome of polyoma virus within its own DNA could transmit the virus to a permissive (murine) host while growing within the animal's intestine. The result was negative; that is, no virus (or tumors) were found.[1]

Although this risk-assessment experiment has often been cited as "evidence" of the safety of recombinant DNA experimentation, it was arguably not particularly well-conceived: a "positive" result (that is, a deleterious effect on a mouse) required a highly improbable sequence of molecular events. This experiment has a present-day analogue: the Silwood study, a complex and elaborate experiment by Crawley et al[2] intended to assess the "invasiveness" of transgenic oilseed rape (canola), is subject to the same criticism, among others. It is a well executed but poorly designed risk assessment experiment.

In three climatically distinct sites and four habitats, the experiment compared the "invasiveness" of three variants of oilseed rape plants over three growing seasons: two varieties modified with recombinant DNA techniques and one "unmodified" variety. The results indicated no important differences in invasiveness among the three varieties.

Although the methodology was elegant and scientifically sound, the result was predictable. It added little to risk assessment research to conclude that transgenic plants with genes for resistance to an antibiotic or to an herbicide are no more invasive than their unmodified parent, in the absence of selection pressure from either the antibiotic or the herbicide in the test environment.

The stated purpose of the experiment was to find out how invasiveness "is affected by genetic engineering." Its motive remains obscure, however, considering the numerous analyses, including those of the NAS and NRC cited above and Crawley's own observation that "...the ecology of genetically engineered organisms is exactly the same as the ecology of any other living thing."[3] Arguably, more useful information would have been derived from exploring the question of how invasiveness is affected by the introduction of certain traits of interest. Even Crawley, the study's director, appears to agree, admitting that if more appropriate and "'risky' constructs like drought-tolerant perennial grasses or insect-resistant weeds had been available, then we would certainly have applied for permission to release them."[4] That better experimental subjects were not available was hardly a suitable rationale for proceeding.

Moreover, the result is not generalizable. The conclusion drawn from the Silwood study, that the particular variants of oilseed rape did not differ in their invasiveness under the conditions tested, enables one to infer nothing about plants in general, or about genetically engineered plants in particular.

Finally, at the risk of splitting hairs, Crawley et al refer to the rDNA-modified oilseed rape as "transgenic,"[5] a term which has come to signify an organism in which genes have been introduced from across species lines by molecular techniques. However, there is evidence that the "unmodified" *Brassica napus*, the species of oilseed rape used in the Silwood study, originated in nature by hybridization between different species or genera. According to Goodman et al,[6] the ancestor of *B. napus* was itself a hybrid between *B. oleracea and B. campestris.* Thus, the distinction between "wildtype" (that is, unmodified) and "genetically engineered" or "transgenic" is not clear-cut, and it is open to question whether the three variants of B. *napus* were sufficiently different to provide a rationale for such a large and expensive ecological study (especially in view of the minimal scientific interest of the "transgenics" that were tested).

The point of this somewhat lengthy digression on the Silwood study is the importance of scientifically valid assumptions. Without them, we do the wrong risk-assessment experiments, hold the wrong conferences and establish the wrong regulatory paradigms (see also chapter 3).

Setting aside the polyoma and Silwood experiments, it is worth considering how one might better go about understanding the risks of recombinant DNA-modified organisms. Two alternative general approaches are discussed in the next essay.

ESSAY 5
THE SEARCH FOR A VALID REGULATORY PARADIGM

An understanding of the risks of recombinant DNA-modified organisms may be acquired in one of two ways. The first is by performing well-designed risk assessment experiments. For plants, for example, such experiments might attempt experimentally to quantify the transformation of a benign, nontoxic, noninvasive plant into one that possesses an undesirable trait(s), or to quantify the ability to induce any plant with a certain newly-acquired trait to transfer that trait by means of outcrossing. An example of the latter would be herbicide resistance under the positive selection pressure of the herbicide in the test environment. However, like those discussed in the previous essay, many such experiments would have a very low probability of a positive result unless they were carefully designed both to maximize the occurrence of a "positive" event and to detect rare events. This approach is cumbersome and often provides data of only limited usefulness. Moreover, as discussed below, it is unnecessary.

Alternatively, one can exploit the consensus view discussed in the previous essay that there is no conceptual distinction between "conventional" (pre-rDNA) genetically altered organisms and those modified with molecular techniques. Based on the appropriate risk assumptions that contributed to that consensus view, the United States National Science Foundation has concluded[1] that risk assessment paradigms and regulatory methods in use before the introduction of the Coordinated Framework[2] provide both a useful foundation and "a systematic means of organizing a variety of relevant knowledge" for the assessment and management of rDNA-modified products. Thus, the rational approach to risk-assessment when risk is not readily demonstrable (a situation sometimes referred to as "very low risk") would use established scientific principles and identify significant gaps in understanding that can be addressed by the conduct of properly designed experiments.

This is a "vertical" approach to understanding risk. It relies heavily on existing knowledge about the behavior of genetic variants produced by nature or human intervention. Thus, a tomato breeder or a government regulator of polio vaccines assessing the potential risks of a new rDNA-derived tomato or vaccine, respectively, is likely to rely more heavily on background information about tomatoes and poliovirus manipulated via traditional techniques than on information about rDNA-manipulated pigs or bacteria.

Important scientific questions relevant to the behavior and performance of organisms in field trials can and should be addressed systematically and in ways that are consistent with recognized scientific principles and procedures. Scientific experience and common sense can suggest approaches that avoid unnecessary and costly experiments performed in the name of "risk assessment." For example, after a field trial of an obviously innocuous rDNA-modified strain of the soil bacterium *Bradyrhizobium japonicum*, Louisiana regulators "felt that continued monitoring of the field [was] needed," and a Louisiana State University professor proposed that this constituted "an opportunity to obtain valuable scientific information from the careful long-term study of the impact of this release on [*sic*] the environment."[15]

The "vertical" approach to risk can also guide effective governmental regulation by drawing attention to the nature of the product (for example, identifying a need for a high level of quality control in the production of airplanes, artificial hearts and nuclear reactors); to circumstances related to risk (such as workplace hazards or railroad cars transporting toxic solvents in populated areas) and, by extension, to existing applicable risk management systems. A vertical approach would take into account, for example, that governmental agencies in the United States, European countries, Japan and certain other countries are experienced regulators. For decades they have overseen the safety of food plants and animals, pharmaceuticals, pesticides and other products that now can be produced using new biotechnology methods. A vertical approach would emphasize the fundamental similarities among products with similar characteristics and the ways that regulators have grouped products or processes in the past, in order to oversee them effectively.

A distorted analytical alternative to the *vertical* approach to regulation and risk assessment is what I have referred to as the *horizontal* approach.[4] In the context of biotechnology, the *horizontal* approach is predicated on the notion that there is something systematically similar and functionally important about the set of organisms whose only common characteristic is genetic manipulation with the techniques of the new biotechnology. It focuses on the artifactual "category" of products made or manipulated with these techniques.

As scientific consensus dictates and the NIH polyoma virus and Silwood risk-assessment experiments have shown, this is not a useful way of approaching or organizing policy toward rDNA-modified organisms, which simply do not comprise a category amenable to generalizations about safety or risk. Nonetheless, the horizontal approach is evidenced in a variety of conferences (discussed in chapter 3) and, as we have seen, risk assessment experiments. At best, the horizontal approach to rDNA technology has been a wasteful nuisance; at worst, it has contributed to specious generalizations and flawed assumptions in policy making.[5]

A horizontal approach to regulatory oversight seldom makes sense, but least so as a basis for crafting governmental regulation. Consider the hypothetical example of widely disparate regulatory requirements for peaches or tomatoes, depending on whether they were mechanically- or hand-picked. In the absence of extenuating circumstances—mechanical pickers spraying toxic oil on the fruit, or workers defecating in the fields during harvesting, for example—dissimilar regulation of the fruit would be hard to justify.

A Paradigm for Biotechnology Regulation

A paradigm for regulating products of the new genetic engineering may be summarized in a syllogism. Industry, government and the public already possess considerable experience with the planned introduction of traditional genetically modified organisms: plants for agriculture and microorganisms for live attenuated vaccines and for other uses such as pest control sewage treatment and mining. Existing regulatory mechanisms have generally protected human health and the environment effectively without stifling industrial innovation. There is no evidence that unique hazards exist either in the use of recombinant DNA techniques or in the movement of genes between unrelated organisms. *Therefore*, for recombinant DNA-manipulated organisms, there is no need for additional regulatory mechanisms to be superimposed on existing regulation. In fact, as a former regulator who was involved for more than a decade with the evaluation of organisms and other products resulting from new genetic engineering techniques, I continue to be impressed with the extraordinary predictability and safety of the newer techniques compared to the older, less sophisticated, "conventional" ones.

The syllogism assumes, of course, that leaving aside the new biotechnology and its products, there exists adequate governmental control over the testing and use of living organisms and their products. This assumption is certainly open to question, particularly where known dangerous pathogens, chemicals and similar products are largely unregulated or where regulations are widely ignored (primarily in developing countries). Nevertheless, throughout our prerecombinant DNA history of scientific research, in the United States and elsewhere, scientists have had a high degree of freedom of experimentation, with pathogens as well as nonpathogens, indoors and out. The resulting harmful incidents have been few, and the benefits, both intellectual and commercial, have far exceeded any detrimental effects.

The syllogism can be illustrated graphically. In Figure 1.1A, the large triangle represents the entire universe of field trials. The horizontal lines divide the universe into classes according to the safety category of the experimental organism (with examples listed on the right side of the figure). These categories can take into account the effect of a genetic change, whether it is a consequence of spontaneous mutation or the use of conventional or new techniques of genetic manipulation. Such genetic changes can cause the organism to be shifted from one safety category to another. For example, if one were to grow mutagenized cultures of *Neisseria gonorrhea* (which causes gonorrhea) or *Legionella pneumophila* (which causes Legionnaire's Disease) in the presence of increasing concentrations of antibiotics to select for antibiotic-resistant mutants, the classification might change from, say, class III to class IV.

Conversely, deletion of the entire botulin toxin gene from *Clostridium botulinum* (the organism that causes botulism) could move the organism from class III to class II or even class I. The oblique lines divide the universe according to the use of various techniques (with techniques becoming newer, moving from left to right).

Consideration of the figure reveals that the use of various techniques does not itself confer safety or risk (except insofar as a genetic change wrought with rDNA techniques is likely to be more precise, better characterized and more predictable). Rather, risk is primarily a function of the *phenotype* of an organism (whether it is wildtype or has traits newly introduced or enhanced in some way) which, in turn, is determined by its genomic information: some organisms are destined to exist symbiotically and innocuously on the roots of legumes (like the bacterium *Rhizobium*) and others to infect and kill mammals at low inoculum concentration (Lassa fever virus).

For an oversight scheme to be "risk based," the use of genetic manipulation techniques generally or certain techniques in particular

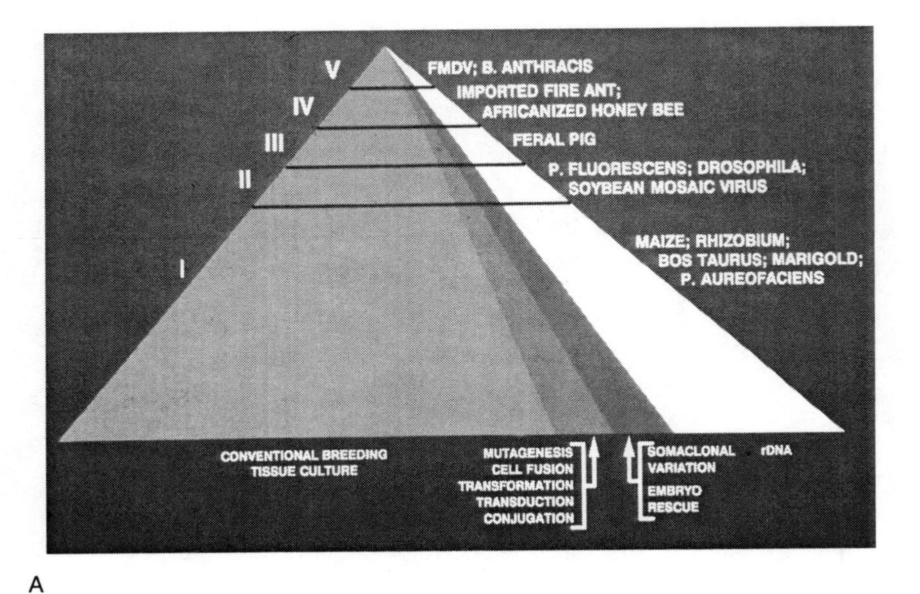

A

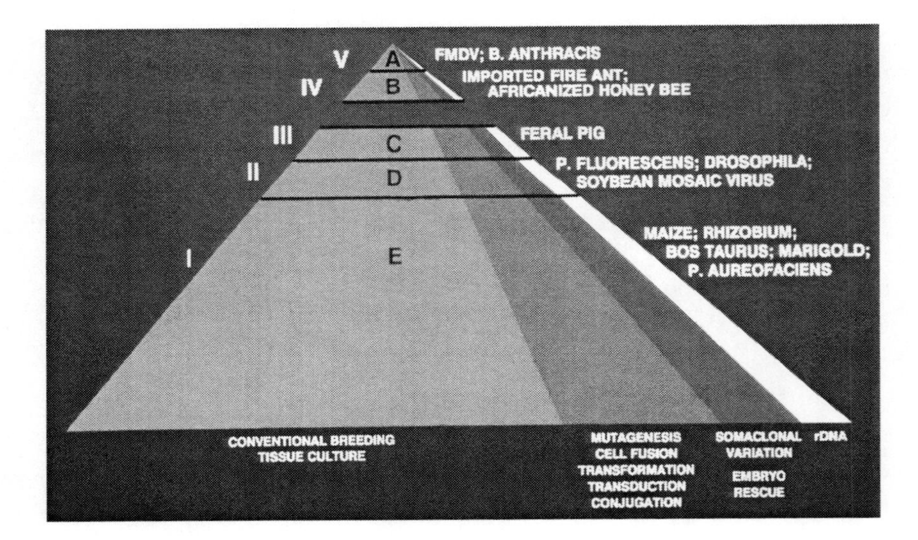

B

Figs. 1.1A,B. The large triangles represent the entire universe of field trials of plants, animals and microbes. The horizontal lines divide the universe into classes according to the safety category of the experimental organism and take into account the nature of a genetic mutation or genetic change. Examples of organisms within the categories are listed on the right. The oblique lines divide the universe according to the use of various techniques (with techniques becoming more recent as one moves from left to right).

should not, per se, dictate the degree of oversight required; rather, the trigger for oversight or for enhanced scrutiny should be a function of the characteristics of an organism—either as wildtype or in any variants. Therefore, there is no coherent scientific rationale for defining the scope of a regulatory net as, say, the "rDNA" slice in Figure 1.1, though that approach has often been proposed.

Figure 1.1B differs from 1.1A by having categories IV and V lifted off to illustrate that this high-risk set of organisms, rather than the "rDNA slice," might be an appropriate "scope" for a regulatory net—particularly if one were designing a regulatory scheme de novo. This suggestion is merely illustrative; depending upon various considerations such as the amount of scrutiny judged to be appropriate and the degree of regulatory burden on researchers and the government felt to be acceptable, the scope of what falls into the "net" could be stratified. For example, category IV and V organisms might be circumscribed for case-by-case, "every-case" reviews by governmental agencies prior to field trials; category III organisms for only a notification; and category I and II organisms exempt. Such a conceptual approach moves us away from discredited technique-based oversight to a more rational consideration of the factors relevant to cost-effective protection from risks. This approach also provides maximum flexibility, in that it can be adapted to more-risk-averse or less-risk-averse regulatory regimes. A risk-based regulatory algorithm that incorporates these principles is proposed in chapter 5.

When Bureaucracy Prevails Over Science

This volume is replete with examples of the scientific consensus on rDNA failing to prevent draconian and regressive governmental responses to the new biotechnology. At the time of this writing, Denmark still prohibits the use of animals that have been genetically modified using the newest techniques, even in the absence of any hint of increased or novel risks to human health or to the environment. For many years, the most innocuous field trials there of a recombinant DNA-modified plant literally required permission from the Queen, which constituted a royal pain for industrial and, especially, academic researchers. Many countries have instituted discrete, stringent regulatory schemes that treat biotechnology R&D in a way that is more restrictive than that for known dangerous pathogens or hazardous chemicals.

Regulatory disincentives are potent. The real-world societal costs of policies derived from invalid scientific assumptions about biotechnology are both large and varied. They include the continued, seasonal loss of crops to frost damage because permission to conduct field trials of the innocuous "ice-minus" bacteria was debated for more than 5 years and the company eventually abandoned the product (see chapter 3); the persistent need for the application of chemical pesticides, while innovative biotechnology-derived alternatives were slowed or abandoned in the face of regulatory uncertainty or hostility; and the sluggish pace of development of rDNA-modified microorganisms for bioremediation, largely because commercial interest has been discouraged by overregulation.

The impacts of the regulatory disincentives will vary widely, in different parts of the world. Columnist George Will has observed that when the *Titanic* steamed into an iceberg, the disaster was not democratic: 56% of third class women passengers died, while only 4 of the 143 first-class women passengers died. You don't need to ask which class was traveling near or below the waterline. When it comes to overregulation of biotechnology, developing countries are likely to be closest to the "waterline."

Developing countries and their vehicle for survival in the next century—namely, food production self-sufficiency—lie in the path of an overregulation iceberg: an iceberg capped by the pale blue headgear of the United Nations. With the advancement of the UN's aspirations to become the world's bio-police, developing countries are in danger of being excluded from the new biotechnology's unquestioned and essential contributions to their agricultural production (chapter 4). Economist Bengt Jonsson is correct that safety can only be bought at a cost.[6] But at *any* cost? Should impoverished countries invest scarce governmental resources in unnecessary environmental assessments of organisms like the ice-minus *Pseudomonas syringae*, extended shelf-life tomatoes and petunias with a novel color?

Regulators remonstrate that they are streamlining regulation and doing better, but that is scant consolation. Biologist and Nobelist Salvador Luria used to say that something that isn't worth doing at all, isn't worth doing well. Moreover, society cannot recover the squandered resources, the wasted time, the abandoned products that past overregulation has cost.

A Caveat for Policy Makers

Princeton physicist and writer Freeman Dyson has observed astutely that resisting a new technology is generally safer for regulators than embracing it, but that this course often has larger costs for society.[7] One cannot attempt to avoid problems with new technology by resisting it or, as the environmentalists seem to want, by trying to stuff the genie back into the bottle—especially if that genie can grant important wishes.

With the accumulated wisdom of the analyses of the NAS, NRC, House of Lords Select Committee and others (*vide supra*), the overriding scientific questions about the new biotechnology have been resolved and the correct assumptions identified. The notions that new biotechnology products defy useful, accurate risk assessment or are too dangerous to introduce into the environment are without merit. To the contrary, it is now clear that there are genuine costs of unjustifiably risk-averse regulatory policies that slow or divert biotechnology innovation and R&D. For these reasons, the professional practitioners and regulators of the new biotechnology must strive to demystify it and to provide an accurate perspective for the public. The stakes are high in terms of both economic and social benefits.

During the past decade or so, the FDA has approved a number of products of new biotechnology that are medical milestones (see Table 1.1). These include α-interferons for the treatment of a lethal leukemia, β-interferon for multiple sclerosis, a monoclonal antibody preparation for preventing the rejection of kidney transplants, new-generation vaccines for the prevention of hepatitis, a hormone that reduces the blood transfusion requirements for kidney dialysis patients, and a growth factor that is an important adjunct to cancer chemotherapy. Among myriad other applications, biotechnology promises improved diagnostic devices for detection of harmful chemicals and pathogens; vaccines against scourges such as malaria, schistosomiasis and AIDS; new therapies that could ameliorate or cure for the first time such genetic diseases as cystic fibrosis and certain inherited immune deficiencies. Used for new generations of medicines and food plants and animals, biotechnology could provide partial solutions to the trinity of despair—hunger, disease and the growing mismatch between material resources and population.

Far from constituting an imminent threat to the environment, application of the new biotechnology may well provide solutions to existing environmental problems. For example, the Asgrow Company developed a virus resistant squash variety that produces tripled yields compared to other varieties and, thus, produces the same overall yield on one-third the farm land. Making crops more efficient through this and other strategies will reduce agricultural land use, freeing it for recreation or for a return to wilderness (and in the developing world, slow the tendency towards up-slope farming). Plants can be biologically protected against pests through genetic engineering, diminishing the need for chemical pesticides at a time when fewer are being registered or re-registered at EPA. Plants can also be engineered to require less fertilizer. Herbicide-resistant plants can offer farmers more weed control choices, favoring the use of herbicides that are less suspect than those used now with some crops. Microorganisms can be engineered to degrade toxic compounds.

Those of us who study the process of government regulation are often asked *why* government regulatory policy and its implementation are so often excessive, poorly-conceived, ponderous, badly implemented, politicized. My colleague, Economics Nobel laureate Milton Friedman, has spent decades studying and writing about this phenomenon. He notes two fundamental—and related—reasons: the influence of special interests, which often induce government to provide substantial benefits to a few while imposing small costs on many; and the self-interest of government officials themselves.

Friedman cites as an example of the first point the licensing of taxicabs in New York City. The number of cabs is limited by government fiat, and this has pushed the price of a medallion (required to operate a taxi) to more than $100,000. The removal of this arbitrary limitation would have wide benefits: consumers would have a wider range of alternatives, the number of cabs and demand for drivers would increase, as would drivers' salaries. So why does the limitation of cabs persist? Because the people who now own the medallions would lose if the limitation were ended. They apply intense economic and political pressure to prevent reform. Conversely, there is no group whose interest in change is large enough to lobby for reform. Calling this phenomenon "rational ignorance,"[8] Friedman observes they are right not to do so.

His second point is that you can usually rely on individuals and institutions (including regulatory agencies) to act in their own self-interest. In the case of regulators, their behavior is influenced both by the desire to avoid errors of commission (even at the cost of committing errors of omission—that is, their discouraging products from being tested or marketed) and by the yearning for larger budgets and bureaucratic empires (Chapters 3 and 5).

Max Planck said more than half a century ago that an important scientific innovation rarely makes its way by gradually winning over and converting its opponents; rather, its opponents gradually die out and the succeeding generation is familiarized with the idea from the beginning.[9] Time is thus on the side of biotechnology in the longer term. There is no doubt that in a decade or two many products made with the new biotechnology will be as much a part of our routines as microwave ovens and computers are to children today.

But that isn't good enough. For many reasons, including those pertaining to economic equity and justice—the poor suffer most from unnecessarily high consumer prices—it would be tragic if its acceptance and potential growth were even minimally limited or delayed. (For those at or below the subsistence level, justice delayed is justice denied.) The greater the unnecessary burden and expense for research and development using biotechnology, the more likely that the earlier biotechnology products will be higher-value-added ones, available predominantly to wealthier countries and populations, and at higher prices than necessary.

The battle for the adoption of sound public policies on biotechnology is not an intellectual exercise akin to a debate at the Oxford Union; rather, it is a real-life struggle for the availability of products that will enhance, prolong and enrich lives—and for the private sector's ability to develop and market such products without undue interference.

It is unfortunate, therefore, to see national and international regulatory policy buffeted hither and yon by political currents. In the United States, since the Clinton administration's installation in January 1993, science and technology policies have been dominated overwhelmingly by a single figure, Vice President Al Gore. These policies, intensely politicized by Gore's singular ideology, have often side-stepped scientific principles and knowledge. The disparity between the Clinton administration's rhetoric about policy and its actions constantly reminds one of Humpty Dumpty's view in *Through the Looking Glass* that a word means whatever he wishes it to mean. The not-altogether-happy tale of U.S. biotechnology policy is treated in the next chapter.

REFERENCES

Essay 1: Biotechnology Policy's Seminal Issue

1. Berg P, Singer MF. The recombinant DNA controversy: 20 years later. Proc Natl Acad Sci 1995; 92:9011-9013.
2. Anon. Coordinated Framework for Regulation of Biotechnology. Federal Register June 26, 1986; 51:23302-23347.
3. Anon. Introduction of Recombinant DNA-Engineered Organisms into the Environment: Key Issues. Washington D.C.: National Academy Press, 1987.
4. Field Testing Genetically Modified Organisms: Framework for Decisions. Washington D.C.: National Academy Press, 1989.
5. Anon. Coordinated Framework, Ibid., emphasis added.

Essay 2: Myths about Biotechnology

1. Anon. Coordinated Framework for Regulation of Biotechnology. Federal Register 1986; 51:23302-23347.
2. Idem.
3. Demain AL, Solomon NA. Industrial microbiology. Sci Amer 1981; 245:66-75.
4. Idem.
5. Anon. Health impact of biotechnology: report of a WHO working group. Swiss Biotech 1985; 2:7-16.
6. Kilbourne ED. Epidemiology of viruses genetically altered by man—predictive principles. In: Fields B, Martin MA, Kamely D, eds. Genetically Altered Viruses and the Environment. Banbury Report 22. Cold Spring Harbor Laboratory, 1985.
7. Klingman DL, Coulson JR. Guidelines for introducing foreign organisms into the U.S.A. for biological control of weeds. Plant Dis 1982; 66(12):1205-1209.
8. Davis BD. Evaluation, epidemiology and recombinant DNA. Science 1976; 193:442.
9. Brisson-Noel A, Arthur M, Courvalin P. Evidence of natural gene transfer from gram-positive cocci to *Escherichia coli.* J Bacteriol 1988; 170:1739.
10. Mazodier P, Petter R, Thompson C. Intergeneric conjugation between *Escherichia coli* and *Streptomyces* species. J Bacteriol 1989; 171:3583.
11. Heinemann JA, Sprague Jr GG. Bacterial conjugative plasmids mobilize DNA transfer between bacteria and yeast. Nature 1989; 340:205.

12. Nester EW, Gordon MP, Amasino RM et al. Crown gall: A molecular and physiological analysis. Annu Rev Plant Physiol 1984; 35:387-413.
13. Takemoto K, Yano M, Akiyama et al. GenoBase 1.1 *Escherichia Coli*, March 1994.
14. Chrispeels M, Sadava D. Plants, Genes, and Agriculture. Boston: Jones and Bartlett, 1994:423.
15. Lincoln DR, Fisher ES, Lambert D et al. Release and containment of microorganisms from applied genetics activities. Report prepared under EPA Grant #R-808317-01, 1983.
16. Anon. Biosafety in microbiological and biomedical laboratories. U.S. Department of Health and Human Services, HHS Publication No. (NIH) 88-8395, 1988, emphasis added.
17. Sharples FE. Regulation of Products from Biotechnology. Science 1987; 235:1329.

Essay 3: Broad Scientific Consensus

1. Anon. Coordinated Framework for Regulation of Biotechnology. Federal Register 1986; 51:23302-23347.
2. Anon. Introduction of Recombinant DNA-Engineered Organisms into the Environment: Key Issues. Washington D.C.: National Academy Press, 1987.
3. Anon. National Biotechnology Policy Board Report. Bethesda, MD: National Institutes of Health. Office of the Director, 1992.
4. Ward M. Do U.K. regulations of GMOs hamper industry? Bio/Technology 1993; 11:1213.
5. Anon. Safety Evaluation of Foods Derived by Modern Biotechnology: Concepts and Principles. Organization for Economic Cooperation and Development. Paris, 1993.
6. Anon. Joint statement from the International Council of Scientific Unions' (ICSU) Scientific Committee on Problems of the Environment (SCOPE) and the Committee on Genetic Experimentation (COGENE). Bellagio, Italy 1987.
7. Workshop Summary. In: Fiksel J, Covello VT, eds. Safety Assurance for Environmental Introductions of Genetically-Engineered Organisms. NATO ASI Series, New York: Springer-Verlag, 1988.
8. Anon. Report of the UNIDO/WHO/UNEP Working Group on Biotechnology Safety. Paris, 1987.

Essay 4: Risk-Assessment Experiments

1. Israel MA, Chan HW, Rowe WP et al. Molecular cloning of Polyoma Virus DNA in *Escherichia Coli*: Plasmid Vector System. Science 1980; 203:883-887.
2. Crawley MJ, Hails RS, Rees M et al. Nature 1993; 363:620-623.
3. Crawley MJ. The Ecology of Genetically Engineered Organisms: Assessing the Environmental Risks. In: Mooney, Bernardi G, eds.

Introduction of Genetically Modified Organisms into the Environment. New York: Wiley & Sons, 1990.
4. Crawley MJ. Arm-chair Risk Assessment. Bio/Technology 1993; 11:1496.
5. Crawley. Nature, Ibid.
6. Goodman RM et al. Gene Transfer in Crop Improvement. Science 1987; 236:48-54.

Essay 5: The Search for a Valid Regulatory Paradigm

1. Covello VT, Fiskel JR eds. The Suitability and Applicability of Risk Assessment Methods for Environmental Applications of Biotechnology. Washington D.C.: U.S. National Science Foundation, 1985.
2. Anon. Coordinated Framework for Regulation of Biotechnology. Federal Register 1986; 51:23302-23347.
3. Anon. Minutes of the Agricultural Biotechnology Research Advisory Committee. Document N. 91-01. Office of Agricultural Biotechnology, U.S. Department of Agriculture 1991; 25-30.
4. Miller HI, Gunary D. Serious Flaws in the Horizontal Approach to Biotechnology Risk. Science 1993; 262:1500-1501.
5. Idem. See also Huttner SL. Government, researchers and activists: the crucial public policy interface. In: Brauer D, ed. BIOTECHNOLOGY. Weinheim, Germany: VCH, 1995:459-494.
6. Jonsson B. The Economic Viewpoint. In: Scientific Innovation in Drug Development: The Impact on Registration. Geneva: IFPMA Symposium, 1986.
7. Dyson FJ. On the hidden costs of saying 'no.' Bull Atomic Sci 1975; 31:23.
8. Friedman M. Why government is the problem. Stanford: Hoover Institution Essays in Public Policy. 1993; 8-9.
9. Planck M. A Scientific Autobiography. New York: Philosophical Library, 1949:33.

CHAPTER 2

The Conflict Between Politics and Science

"Although everybody with a political agenda routinely professes great respect for the wisdom of the American people, the actual behaviors reveal not respect but thinly veiled contempt."‡

ESSAY 1
INTRODUCTION

On the main building of the Massachusetts Institute of Technology stands a frieze honoring the immortals of science—Newton, Archimedes and Pasteur, among others. It reminds those who enter what science and its practitioners can achieve. Somewhere there should also be a monument dedicated to those who have *impeded* scientific progress—a troglodyte frieze. It would immortalize the persecutors of Galileo and the supporters of Soviet Lysenkoism. And those of us who have served as midwives at the birth of the new biotechnology could add a few others. The list would include the overzealous souls who push for governmental overregulation in the name of consumer or environmental protection, the sophists who spark fear in the heart of the naïf with chilling characterizations of "Andromeda Strains," "eugenics" and "Frankenfood." Their seemingly endless assaults have compelled some of us in the scientific community to fight back. The essays in this section represent some of my forays into the battle over the future of science and technology.

---

‡Anderson WT. Reality Isn't What it Used to Be: Theatrical Politics, Ready-to-Wear Religion, Global Myths, Primitive Chic, and Other Wonders of the Postmodern World. New York: Harper & Row, 1993.

I have been wrong consistently about the durability of the product versus process debate. In the early 1980s, I thought the issue was put to rest when Professors Allan Campbell and David Baltimore (eminent molecular biologists and members, at that time, of the NIH Recombinant DNA Advisory Committee (RAC)) proposed that the NIH rDNA Guidelines become voluntary. This would have been an important advance. It would have acknowledged, at least implicitly, that the Guidelines had been predicated on the incorrect assumption that rDNA techniques, among all methods of genetic manipulation, were somehow different with respect to risk. Already, an impressive body of experimental data—as well as sober reflection—had made a compelling case for change.

At the time, I was the FDA representative on the RAC (and hold the distinction of having served longer (1980-1993) than anyone before or since). I supported the Baltimore-Campbell reform effort. The majority of RAC members expressed support for the general arguments about the nature of rDNA risks, and they agreed that rDNA methods and products warranted no special precautions or oversight beyond that applied to standard microbiological practices. Then they proceeded to reject the proposal.

These same committee members stated pointedly that the reforms would unacceptably diminish the RAC's role. They believed that the RAC meetings served a valuable public relations purpose. The open meetings, they argued, provided a prestigious and democratic forum in which anyone could air concerns about the testing and use of new biotechnology methods and products. Although their ostensible goal was to gain public approbation of scientific research, some members revealed an unmistakable element of self-congratulation and self-importance.

These attitudes are not uncommon. In part, they reflect the collective angst of a scientific community haunted (inappropriately, in my view) by an earlier generation's Manhattan Project and remembering the socially conscious and oh-so-sensitive 1960s. They also reflect professional vanity. Who wants to characterize as superfluous something to which he has contributed?

Unfortunately, vanity has its price. In sustaining the rDNA Guidelines, the RAC sent a signal that NIH and the scientific community were still concerned about the new genetic methods.

I was slow to learn. I doubted, perhaps more strongly than ever, that the product versus process debate could survive a more strenuous scientific analysis. I was wrong. The definitive, 1987 National Academy of Sciences white paper[1] (chapter 1) and all of its cogent findings on rDNA risk went largely unheeded. USDA and EPA rolled out scientifically indefensible regulatory schemes that set new standards for federal agencies' aversiveness to (negligible) risk and to novelty.

Just as with the NIH rDNA Guidelines, these ill-conceived policy efforts were abetted by agency staff, academic scientists and company executives who sought some ill-defined "public" approval. In the process, of course, they also aggrandized themselves and their institutions by creating an aura of civic-mindedness.

The muted retort of these regressive policy decisions sounded through the decade of the 1980s like a Chinese firecracker caught in slow motion. Federal bureaucrats eager to expand their regulatory domains, companies desperate to calm jittery investors and academic scientists eager for public support constituted the packets of gunpowder. The sensationalized claims of antibiotechnology groups—promulgated by the media, hungry for lurid scenarios—lit the fuse.

I have consistently underestimated the degree of tolerance that would be extended to the activists and their mythic visions of apocalypse. The unflagging efforts of professional agitators like the Union of Concerned Scientists' Margaret Mellon (formerly of the National Wildlife Federation), the Environmental Defense Fund's Rebecca Goldburg, the Foundation on Economic Trends' Jeremy Rifkin and Consumers Union's Michael Hansen have been given considerable access and credibility in many quarters. These quarters include Gregory Simon (domestic policy advisor to Vice President Al Gore and former congressional staffer); the USDA and EPA staff who weigh equally activist visions and scientific judgments; certain members of the House and Senate and their staffs; and individuals in companies and universities who foolhardidly hope to win the activists over, or buy them off.

The antibiotechnology activists discard like yesterday's newspaper the considered findings of distinguished scientific organizations. They claim that there is no documentation, just opinions, on matters of rDNA risk. To my astonishment, their endless plaint was met in the 1980s with a half-million dollar study by the National Research Council (NRC), funded by several federal agencies.

That the resulting NRC report[2] (chapter 1) wholly confirmed the conclusions and recommendations of the earlier NAS report was no surprise. It was similarly no surprise that the report was shrugged off, yet again, by the troglodytes. It was, however, both surprising and inexcusable that EPA and USDA were allowed to proceed with their innovation-averse schemes by a community no less broad and influential than the federal government, Congress, industry and academic researchers. These new regulatory proposals set precedents for treating genetically modified organisms as especially dangerous—because they are modified with our newest, most sophisticated and most precise genetic techniques. The agencies' hubris flew in the face of decades of positive experience with genetically modified plants, animals and microorganisms in the absence of governmental regulation.

I must also admit that I was unaware of the magnitude of financial resources available to the activists. What appears superficially to be an inconsequential handful of ideologues is actually a collection of well-financed professionals employed by organizations to pursue single-issue activism. The same handful have battled biotechnology over two decades, winning occasional (modest) successes. They appear at public hearings claiming to represent hundreds of thousands of concerned citizens, although not one of their organizations has undertaken any grassroots assessment to support the claim. I have wondered, for example, how the rank-and-file members of the National Wildlife Federation would have reacted to the organization's bitter opposition to the development of a rabies vaccine for animals in the wild, had they known about it.

The antibiotechnology professional activists have promoted a pseudocontroversy marked by surrogate issues which have ranged from wholly imaginary safety concerns to inaccurate economic forecasts for small farms and trumped-up consumers' rights. They have claimed that even the most modest, precise and well-characterized genetic modification can have unpredictable, disastrous effects. They have said that using bovine somatotropin (bST) to increase milk production in dairy cows will cause breast cancer in women who drink milk. They predicted that field trials of the "ice minus" bacterium *Pseudomonas syringae* could disrupt weather patterns and air traffic control. The broadest claims are that consumers have a right to know via product labels about all techniques, materials and sources that are used in making any food—and that all rDNA-derived foods should be clinically tested.

Antibiotechnology antics have ranged from the silly to the bizarre and pernicious. In the 1970s a group of activists disrupted a scientific conference chanting "we shall not be cloned." Strawberry plants being tested in field trials have been ripped from the ground. A few hard core activists have campaigned relentlessly against all biotechnology applications, including production of pharmaceuticals for treating and preventing cancer, heart attacks, AIDS, hepatitis and rabies. They continue to beleaguer government agencies with petitions to publicize, restrict or altogether outlaw new biotechnology products and research. One campaign, completely baseless but nevertheless well-orchestrated, publicly accused then-FDA Deputy Commissioner Michael Taylor of a conflict-of-interest related to FDA's biotechnology policy formulation and product evaluation.

I have underestimated consistently how easily these critics could capitalize on not only the public's scientific naivete, but on *others'* concern about the public's scientific naivete. The critics successfully gambled that government regulators, industry executives and university scientists would panic over the *possibility* that the public would be taken in by the antibiotechnology controversies. Unfortunately, little attention was paid to public opinion polls that showed that the public was not up in arms about biotechnology. In an odd twist on George Orwell's observation about the public's "vague fears and horrible imaginings," the activists mined a vein of leftist puritanism.

The overall theme of antibiotechnology activism is not altogether new. It resonates well with historian Richard Hofstadter's classic analysis of religious and political movements in American politics. Hofstadter described the religious and political activists' obsession as the "paranoid style," summarizing it this way: "The central image is that of a vast and sinister conspiracy, a gigantic and yet subtle machinery of influence set in motion to undermine and destroy a way of life." He goes on to note a characteristic "leap in imagination that is always made at some critical point in the recital of events."[3]

Susanne Huttner, director of the University of California Systemwide Biotechnology Research and Education Program, has placed the biotechnology critics squarely in Hofstadter's sights:

viewed from Hofstadter's model of the paranoid style, the "conspiracy" here lies in industrial agriculture, and the "leap in imagination" lies in the assertion that biotechnology is at base bad for agriculture, farmers and developing nations.[4] All biotechnologies? What about plants resistant to disease, insects or drought? Veterinary diagnostics and vaccines? Grains with enhanced nutrient content?

There is a consequential lesson to be learned from all this. Antibiotechnology groups are entitled to their own views of science and technology, but it is foolish to delegate to them responsibility for defining the public interest or for estimating the potential risks or benefits from basic research and product development. The scientific community, spanning government, industry and academia, should not be sanguine about activist campaigns on behalf of "the public" when the basic tenets of those campaigns contradict empirical knowledge and defy scientific principles of risk analysis.

Consider, for example, Greenpeace International, which may have attained the nadir of antibiotechnology activism when, on April 6, 1995 the organization announced that it had "intercepted a package containing rice seed genetically manipulated to produce a toxic insecticide, as it was being exported...[and] swapped the genetically manipulated seed with normal rice."[5]

Here, we have the apparent result of too little education and morality combined with too much Mission Impossible on the telly. The rice seeds stolen by Greenpeace had been genetically improved for insect resistance and were en route to the International Rice Research Institute in the Phillippines from the Swiss Federal Institute of Technology in Zurich. The modified seeds were to be tested to confirm that they would grow and produce high yields of rice with less chemical pesticide. In the Phillippines and many developing countries in Asia where rice is a staple food, disease-resistant and insect-resistant rice are desperately needed.

Greenpeace interfered with this research project only because the seeds had been genetically modified using rDNA techniques. Each year, thousands of other genetically modified seed samples are shipped to and from the International Rice Research Institute and other agricultural research centers around the world—without notice by Greenpeace. Mindless of their folly (and no doubt breathless from their adventure), Greenpeace proclaimed:

> "Transports of hazardous waste have to be approved by both export and import countries as well as by all transit countries along the way. In Switzerland, the Federal Office of the Environment (BUWAL) watches over the exports of toxic wastes. Unlike chemical substances, genetically engineered organisms have the potential to multiply, spread and simply get out of control. Obviously such organisms constitute a danger for people and for the environment....It should be clear that the export of genetically manipulated organisms needs to be even more tightly regulated than the export of toxic wastes."[6]

If Greenpeace were to prevail, the exchange of plant germplasm and the enhancement of crops for indigenous farmers would cease to exist. Who will underwrite the costs of regulation that is, without any justification, more stringent than that for "export of toxic wastes?" New technologies and products would become sequestered among industrialized countries, whose populations could bear the inflated costs of overregulated consumer products. What—and whose—public interest is Greenpeace serving?

This is a worst-case example of how antitechnology groups intentionally blur distinctions between products, like seeds and toxic wastes, in order to elicit the most stringent regulatory requirements possible. A new tomato variety is not and should not be regulated as stringently as a new drug or vaccine. Clinical trials are routinely—and appropriately—required for new pharmaceuticals, and rightly so. But professional standards adopted voluntarily by plant breeders have been used routinely and effectively for decades to test new varieties of fruits, vegetables and grains. Plant breeders test yield, flavor, toxicity and properties such as resistance to insects, mold and viruses. Unsafe or otherwise unacceptable plants are discarded from breeding and barred from commerce.

Plant breeding has achieved an impressive record of safety, based on professional, rather than government-imposed, standards of practice. Choosing to ignore this record, Greenpeace, the Environmental Defense Fund and other antibiotechnology groups call for extensive environmental and human testing of new agricultural biotechnology products. They know very well that profit margins for fresh and processed foods are extremely narrow and the markets are highly competitive. They are counting on the new regulations to add to the cost of developing new biotechnology products, effectively moving them out of the reach of farmers, markets and consumers. While I have learned much from the product versus process debate, I will never understand what they hope to gain by keeping improved staple crops out of the hands of farmers, sustaining their reliance on high acreage, input-intensive farming methods.

Far too often, government policy makers have welcomed antitechnology activists to their advisory committees, hearings, conferences and bosoms. Biologist Donald Kennedy, former FDA Commissioner and Stanford University president, has analyzed various aspects of governmental oversight of America's scientific enterprise. Bringing to it the experience of a scientist and regulator, Kennedy observes that bad public policies usually result when we respond politically to some popular movement, only to discover that we have mistaken its real motivation. "'We did what they wanted, but after we did it they turned out to want something else' is among the oldest of political complaints. It has all kinds of bad consequences. Not only is the wrong policy put in place, but those who have tried to be responsive experience alienation and disillusionment when they discover that they have not provided any satisfaction."[7]

Kennedy gently chides policy makers: "Frequently decisionmakers give up the difficult task of finding out where the weight of scientific opinion lies, and instead attach equal value to each side in an effort to approximate fairness. In this way extraordinary opinions, even those like Mr. Rifkin's, are promoted to a form of respectability that approaches equal status."[8]

Kennedy is too charitable. In the biotechnology arena, the policy makers have used the high profile demands of antiscience groups to justify regulatory nostrums.

While science and science policymaking thrive in the marketplace of ideas, we must differentiate between science and pseudoscience. Organizers of NIH-sponsored conferences on genetics do not invite creationists to revisit the validity of basic tenets of evolutionary biology. Applied physics meetings do not include sessions devoted to the newest designs for perpetual-motion machines.

To my well-meaning colleagues in academia, government, industry and nonprofit organizations who would attempt to carry on meaningful dialogue with the antibiotechnology activists, I have but one word: *Don't.* Their agenda is simply to arrogate control over what research is performed, what tools are used and what products are brought to market. That agenda cannot be advanced by scientifically reasonable arguments, or by acknowledging the primacy of empirical evidence or the scientific method. There is little common ground.

For their steadfastness and zeal, the antibiotechnology groups genuinely deserve a monument. To capture the essence of their battlefields and tactics, perhaps this "troglodyte frieze" should depict the activists borne on the shoulders of federal bureaucrats.

ESSAY 2
YOU CALL *THIS* A PRO-TECHNOLOGY POLICY?

The actions of U.S. presidential administrations in this century have often not matched their words. Wilson and FDR pledged they would keep the United States out of wars. Reagan promised he would eliminate the federal budget deficit, and then there was Bush's "Read my lips, no new taxes." Clinton's "end of the era of big government" in the 1996 State of the Union speech also belongs on the list. (When journalists followed up with enquiries as to which federal departments he intended to close, the president answered boldly, "We plan to eliminate many programs, such as the Tea Tasters' Board, Uniformed Services University of the Health Sciences and the Naval Academy's own dairy farm.")[1]

There are other notable examples of the Clinton administration's disconnect between words and actions. Nowhere has this phenomenon been more evident than in science and technology policy. Government officials have extolled the importance of science and technology to the economy, jobs and quality of life, while their actions consistently tell a different story. One need only contrast the palpably negative impacts of their biotechnology policies with the lofty goals outlined in a slick administration booklet, "Science and the National Interest," published in August 1994.[2]

Consider several examples. Goal: "enhance connections between fundamental research and national goals," including "stable policies on research funding." In fact, public sector funding is down. Budgets for 10 biomedical institutes of the National Institutes of Health were on the administration's cost-cutting block for fiscal year 1995; overall, as a percentage of the nation's gross domestic product, the White House's proposed science budget for that year was the lowest since 1958. Only the concerted action of the congressional Republicans provided a solid increase for NIH in FY96—a whopping 5.7% boost from the previous year, $175 million more than the administration's request.

While Congress has generally supported funding of investigator-initiated academic research, the president repeatedly emphasizes the federal government's own "critical role to play" in performing and directing research and development. Moreover, he derides the budget cutters when they find needed funding by trimming ineffective technology programs at the Department of Commerce and the Environmental Protection Agency.

While small increases in Mr. Clinton's FY97 budget appear to keep National Institutes of Health and the National Science Foundation ahead of inflation, buried in the fine print are projections of sizable reductions in extramural grants, and in many science and technology programs at other agencies. For example, excluding the funding earmarked for refurbishing the NIH's on-campus hospital, its actual budget increase is only 1.3%, well below the rate of inflation.

The closer one looks, the more dire the implications. Clinton administration largess is targeted to the agencies that historically have funded the lowest quality and most applied research. For example, some of the largest proposed increases are for the Department of Commerce (up 16.9%, to $4.3 billion) and the EPA (up 22.8%, to $7 billion). (In 1995, the EPA was harshly criticized by a National Academy of Sciences panel for low scientific standards and lack of peer review of its research.) The Clinton-Gore policies seriously worsen the plight of American universities, plagued by shrinking budgets, aging research facilities and an unwillingness by the federal government to pay reasonable indirect research costs.

Goal: "stimulate partnerships that promote investments in fundamental science and engineering and effective use of physical, human and financial resources." The reality is that biotech was caught both directly and indirectly by the White House assaults on drug prices and physicians' prescribing, and its insistence on rebates for Medicare and Medicaid drugs. Not surprisingly, small biotech firms (often without products on the market or any other source of revenues) were harder hit by the defection of potential investors than pharmaceutical giants with long records of product development and marketing. During the Clinton administration, the former have seen investment capital from public offerings fall and stock market valuations underperform other sectors of high-tech stocks.

In theory, these data could simply be part of a normal business cycle: healthy companies spending the capital they raised in earlier years. However, that interpretation is not consistent with biotech entrepreneurs' complaints that, for a time, venture capitalists stopped returning their phone calls. A survey by the Biotechnology Industry Organization of companies involved in AIDS research also pointed to the administration's culpability. Forty percent of the companies surveyed cited White House "cost containment" proposals as the cause of the industry-wide shortage of capital. Moreover, a survey by the Gordon Public Policy Center at Brandeis University found that of companies that planned to raise capital in 1994, 60% said they either fell short of their funding target or withdrew or postponed stock offerings.

Adding insult to injury, the Clinton White House killed the Biotechnology Presidential Initiative that grew out of the previous administration's extensive analysis of areas ripe for supplemental funding. The initiative could have capitalized on U.S. strengths, remedied competitive weaknesses, and catapulted American biotechnology farther ahead.

Goal: "a stable science-based regulatory system." The Clinton-Gore report asserts that the administration has "taken significant steps...towards accelerating the development of technologies critical for long-term economic growth and for increasing productivity while reducing environmental impact...[via] fundamental science." This is the big lie. The reality is that at the same time that funding has been tenuous, biotechnology regulation, the *bête noire* of the administration's policies, has been increasingly anti-innovative, unscientific and focused on negligible-risk activities. This is discussed at length in subsequent chapters.

The Department of Agriculture (USDA) has required unnecessary permits for more than 1400 field trials of genetically modified plants, all of them of negligible risk. Even after the Congress attempted to kill off the USDA's Office of Agricultural Biotechnology, a paragon of superfluity and meddlesomeness, an undersecretary sought ways to preserve some of its functions. The

Food and Drug Administration (FDA), which has a generally positive 15 year track record on regulating biotech, has begun to revise its regulatory approach as instructed by the administration's political minions.

The Environmental Protection Agency's (EPA) biotech regulatory policies—and its attitude (in both the traditional and colloquial senses)—are perhaps the most egregious of all (chapter 3). In September 1994 the EPA published regulations for biotech biocontrol agents. In these regulations, the EPA targeted only products made with the most precise and predictable new genetic methodologies.

In a proposed regulation published in November 1994 the agency made known its plans to expand its regulatory dominion to a whole new category of products—whole plants that are made resistant to pests by using the new genetic techniques. If the EPA has its way, these garden and farm plants will be regulated even *more stringently* than chemicals similar to arin, DDT or parathion. Yet American plant breeders have been creating genetically improved plant varieties, very often with pest- or disease-resistance as the goal, for more than a century without government regulation.

The essence of the EPA's policies is that the use of certain techniques—rather than considerations related to risk—triggers regulation. Under these policies, phenotypically identical organisms may be regulated very differently, if different genetic techniques were used in their construction. In the product versus process debate (discussed in chapter 1) the EPA comes down solidly on the process side, contrary to the consensus of the scientific community.

These regulations make neither economic nor scientific sense. The USDA, FDA and EPA regulatory approaches constitute, in effect, a tax on innovation that uses the new biotechnology. Ineluctably, these anti-innovative policies will discourage research using the newer, more precise techniques, denying consumers more varied and nutritious produce grown with fewer chemicals. Such policies have been especially burdensome to the academic research community, which, unlike industry, cannot simply write off excessive regulation or absorb it as a cost of doing business. (Most research grants for academic investigators working on agricultural biotechnology are only in the tens of thousands of dollars.) Too many university researchers have been doing paperwork instead of experiments.

In the Clinton administration, Vice President Gore is the General leading the regulatory agencies' little known campaign against science and technology. Commissioned to the task by executive order, he is not marching alone however. The administration has put in place a staff of administrators and advisors who are shockingly callow and shallow.

For example, Ellen Haas, former director of an antitechnology advocacy group and lately promoted from Assistant Secretary to Undersecretary of Agriculture, authored an article that demonstrates little comprehension of science's ability to provide objective findings about food safety. "You can have 'your' science or 'my' science or 'somebody else's' science," she wrote. "By nature, there is going to be a difference."[3] (That comes as a revelation to someone like me who has been publishing papers in and reviewing manuscripts for peer-reviewed journals for upwards of two decades.) Moreover, she sees food safety issues as inevitably involving a polemic: "you have, on one hand, the economic interest of the food industry, and on the other, the health interest of consumers."[4] In her new subcabinet-level position, Haas now brings these cynical and antiscience views to the highest levels of government policymaking.

It seems self-evident that the president's science advisor should be an advocate for science and technology and for policies that promote research and development. However, the biotechnology record of Jack Gibbons, Clinton's science advisor, is dubious. As the director of the congressional Office of Technology Assessment, he signed and vociferously defended a 1988 report that served as a virtual L.L. Bean catalogue for antibiotechnology ideas and policies. *New Developments in Biotechnology—Field Testing Engineered Organisms*[5] was nothing more than ideology masquerading as science. It may be the least scholarly and accurate analysis of a biotechnology issue of the dozens produced by the U.S. government during the past two decades. Certainly, during his tenure as a senior advisor to the president, no one has accused Gibbons of advocating science as the basis for regulatory policy—or, in fact, of doing much of anything. The ultimate goal of the White House's Office of Science and Technology Policy under Gibbons seems to be to convene advisory panels and to produce reports that are a mile wide and an Angstrom deep.

FDA Commissioner David Kessler's contributions to biotechnology have been even less auspicious. They include: (1) The elimination of the two FDA policy offices with extensive involvement with the industry (the Office of Biotechnology and the Office of Small Business, Scientific and Trade Affairs); (2) Presiding over a three year review of a long-shelf-life tomato (by contrast, the reviews of the first two new biotech therapeutics in the early 1980s—human insulin and human growth hormone—required 5 and 11 *months*, respectively); (3) Directing the preparation of a new notification requirement solely for new biotech foods, reversing 15 years of risk-based policies; (4) Instituting a vigorous FDA-wide search for reasons *not* to approve bST (bovine somatotropin, a protein that increases the productivity of dairy cows), on instructions from the administration. (Moreover, now that bST has been approved (reluctantly), the FDA is permitting retailers to imply incorrectly in labeling that there is a material difference between milk from bST-treated and untreated dairy cows, which has the effect of "damaging" the acceptability of both bST and the milk from treated animals) and (5) During Congressional testimony, implicating the new biotechnology—inaccurately, as it turned out—in the creation of a high-nicotine tobacco plant.

The Clinton administration's opposition to bST, apparently derived from a desire of Vice President Gore and his staff to "teach industry a lesson" for daring to develop such a politically incorrect product (it's a hormone that "tampers" with milk allegedly threatens small family farms etc., etc.), is particularly ironic. Why ironic? The answer has several parts. BST is a product that increases dairy cows' productivity and favors those farms that are well-managed, regardless of size. And consider this excerpt from Hillary Clinton's *It Takes a Village* (Simon & Schuster, 1995):

> I will never forget the woman from Vermont whom I met at a health care forum in Boston. She ran a dairy farm with her husband, which meant that she was required by law to immunize her cattle against disease. But she could not afford to get her preschoolers inoculated as well. "The cattle on my dairy farm right now," she said, "are receiving better health care than my children."[6]

The First Lady and the vice president need to have someone explain in very simple terms the direct correlation that exists between government policies that encourage product innovation and citizens' well-being: if government agencies keep the regulation of research and development to the level that is necessary and sufficient, the quest for profits stimulates researchers' and industry's interest in making products like bST. Farmers then use bST, which increases the productivity of their cows roughly 10-20%. This, in turn, enables them to produce the same amount of milk with fewer cows, milking machines, veterinarians' visits, impregnations and inoculations. The farmers make more money and are able to afford to have their children vaccinated. (But the Clinton administration has *bashed* bST—as well as other manifestations of biotechnology—at every opportunity.)

Let us return to the FDA under David Kessler. Previous FDA Commissioners (five of whom I worked for) had insisted that agency decisions be based on science, law and common sense. Now everything has become politics, politics and more politics.

Even decision-making on individual FDA-regulated products has become politicized. Jerold Mande, the agency's political commissar and a former Gore staffer, is making key regulatory—and even some civil service personnel—decisions (see chapter 3). Even more destructive is Gore adviser Gregory Simon.

Simon, Gore's senior domestic policy adviser, a lawyer without scientific training or experience (and, therefore, a typical Gore choice for directing science policy), has long been a nemesis to the new biotechnology. While a staffer on the House Science, Space and Technology Committee he authored the Biotechnology Omnibus Act of 1990, HR 5232. This notorious proposal for comprehensive regulation of field research foundered in committee. Had it passed, however, HR 5232 would have established a vast regulatory infrastructure focused exclusively on organisms manipulated with precise molecular "gene-splicing" techniques. We would have had a U.S. statute squarely at odds with the scientific consensus described in chapter 1.

The bill, which was science-averse without even a pretense of sheltering consumers from genuine risks, would have created potent regulatory *dis*incentives to the use of the most precise and best genetic technology. Such a development could have begun the devastation of the biotech industry several years before the

Clinton administration's health care reform and regulatory policies attempted it in the mid-1990s.

Since becoming the vice president's aide, Simon's speeches have reflected his antagonism toward and lack of understanding of technology. Simon has said that the actual degree of biotech's risks is irrelevant, that the new biotechnology must be subjected to a high degree of governmental control and regulation in order to calm a "hysterical" public. In other forums, Simon has amplified his message, saying that, for regulatory purposes, biotech products simply cannot be compared to traditional products and that "consumers will have to change their concept of how food is made" before they will accept the technology.[7]

It is difficult, however, to reconcile such assertions with consumers' actual attitudes and behavior: surveys indicate that consumers are optimistic about the fruits of the new biotechnology and surprisingly sanguine about its risks. Furthermore, new-biotech-produced chymosin, an enzyme approved by the FDA in 1990, is currently being used for the manufacture of more than 60% of the cheese made in the United States. Several vegetables produced with rDNA techniques are also on the market and bST is used in more than 15% of the nation's dairy cows, with nary a complaint from consumers and without special labeling for any of these products.

Simon's views are diametrically opposite to the unequivocal conclusions and recommendations of the world's experts. More important, they are *not* pro-consumer as he would like to claim. On the contrary, his policies deprive consumers of the availability of new products; in effect, impose a tax on the use of new technology, making products more expensive; and condescend to consumers by misinforming them about risk. Rather than reassuring consumers about the safety of a new technology and its products, excessive government oversight conveys the opposite message. As the president of an advocacy group, Consumer Alert, testified, "For obvious reasons, the consumer views the technologies that are *most* regulated to be the *least* safe ones. Heavy involvement by government, no matter how well intended, inevitably sends the wrong signals. Rather than ensuring confidence, it raises suspicion and doubt."[8]

What could be the motivation for the antitechnology actions of those who extol the importance of technology? Vice President

Gore's attitudes, ascertained from his extensive writings over many years, provide a clue.

While a congressman and self-styled expert on biotechnology issues, Al Gore praised Jeremy Rifkin's shoddy antibiotechnology diatribe, *Algeny*, as "an important book" and an "insightful critique of the changing way in which mankind views nature."[9] In a 1991 article in the *Harvard Journal of Law and Technology*, then-Senator Gore displayed an astounding lack of appreciation of the historically positive linkage between science and economic development, when he described investors' eager reception of Genentech's 1980 stock offering disdainfully as the first sellout of the "tree of knowledge to Wall Street."[10] Even before Genentech and gene-splicing, biotechnology had yielded impressive contributions, using microorganisms to produce antibiotics, enzymes, vaccines, foods, beverages and other products. Biotechnology was a $100 billion industry long before the gene-splicing techniques emerged. Nevertheless, undeterred by—or ignorant of—this record, Gore claimed that "the decisions to develop ice-minus [bacteria], herbicide-resistant plants and bovine growth hormone...*lent credibility to those who argued that biotechnology would make things worse before it made things better*" (emphasis added).[11] One has to marvel at the vice-president's vision of how a harmless and ubiquitous bacterium that prevents frost damage to crops, or plants that will reduce the use of agricultural chemicals and provide farmers with new tools, could be detrimental.

In the same 1991 article, Gore coins a "principle that applies to regulating new and strange technologies such as biotechnology," in order to rationalize unnecessary congressionally-mandated regulations for biotechnology: "If *you* don't do it, you know somebody else will." (emphasis in original).

In an original but bizarre twist, Gore worries even about biotechnology's possible success:

> The most lasting impact of biotechnology on the food supply may come not from something going wrong, but from all going right. My biggest fear is not that by accident we will set loose some genetically defective Andromeda strain. Given our past record in dealing with agriculture, we're far more likely to accidentally drown ourselves in a sea of excess grain.[13]

I doubt that that apprehension is shared by developing countries, confronted by the prediction that over the next century the world's population is expected to more than double, from 5.5 billion to about 11.3 billion people, with more than 80% of the additions expected to reside in their regions.[14]

Gore's attitudes toward biotechnology are consistent with his negative view of science generally, which he bares in *Earth in the Balance.* As discussed in the following essay, his book provides a disturbing insight into the thinking of America's technology czar. Throughout, Gore employs the metaphor that those who believe in technological advances are as sinister, and polluters are as evil, as the perpetrators of the World War II Holocaust. These views—expressed repeatedly over many years, in many forums and contexts—cannot be reconciled with the Clinton administration's self-congratulatory, upbeat rhetoric and the endless technology-related "photo-ops."

ESSAY 3
POWERFUL IDIOCY: LYSENKO, GORE AND U.S. BIOTECHNOLOGY POLICY

It has been said that those who do not remember the past are condemned to repeat it. It is important, therefore, to consider the parallels between the decimation of basic and applied biology in the Soviet Union earlier in this century and the battering of present-day biotechnology by the Clinton administration and its allies in the "green movement." In both cases, we see the sacrifice of new science to old myth; heterodox, unscientific theories steering public policy; the abject failure of that public policy, with dire outcomes for research and commerce; and glib, condescending exclusionary attitudes toward policymaking.

Some may think it unlikely that there would be instructive parallels in policymaking between such different nations and eras, but there is compelling evidence that the Clinton administration is driven by a philosophical orientation hostile to new science. What flows from that philosophical orientation is aberrant policymaking for science and technology. And that is a cause for worry because it is barely 30 years since the world witnessed the fall of a tyrant who had devastated science in the Soviet Union.

LYSENKO

The 20th century's "*enfant terrible*" of science policy is, without question, Trofim Denisovich Lysenko. Lysenko ravaged basic and applied biology in the Soviet Union in a reign of terror that lasted from the mid-1930s until the mid-1960s.

Lysenko and his minions seized control of the institutional centers of Soviet agricultural science at an opportune moment. A 10 year crisis in grain production had left bureaucrats yearning for quick fixes. Lysenko had them, proposing easy ways to increase the productivity of plants, ways that suited the Communist agenda.

Lysenko challenged the dogma that genes convey hereditary traits, preaching instead Lamarckian theory (after French naturalist J.B. de Lamarck, 1744-1829): that environmental influences cause structural changes in animals and plants that are transmitted to offspring. Lamarckian theory had superficial plausibility. (Did it not seem plausible that giraffes would stretch their necks to reach tender shoots on high tree limbs, and that in this way the trait of an elongated neck would be passed on to future generations?) More important, the political correctness and utility of Lysenko's theories were of particular appeal to his masters: standard evolution would not produce the desired results rapidly enough, but the inheritability of acquired characteristics provided a speedier path. This reasoning was applied both to the creation of the New Soviet Man and to agriculture. The new social order could even conquer Nature by expediting improvements. A paragon of Soviet science and political ideology, Lysenko received the full imprimatur of Stalin and the Party.

According to the author on international politics Robert Conquest, "Lysenko owed his long triumph, almost equally damaging to agriculture and to science (and scientists), to having as his main supporter a man skilled in the manipulation of Marxist terminology, I.I. Prezent, who had no difficulty in proving genetics reactionary and its practitioners enemies of the people."[1]

Lysenko's success shows that the improbable can, under the right confluence of circumstances, become the basis for critical policies. Here was a meagerly educated agronomist and technocrat who claimed, for example, that wheat plants cultivated in the appropriate environment produce seeds of rye. This is tantamount to saying that dogs living in the wild give birth to foxes. The remedies for agriculture he proposed were destined to be exercises in futility. Yet, with the support of Stalin and the Party, he became an irresistible force.

Via persecution of dissenting scientists and political control of science, Lysenko (as director of the Institute of Genetics of the Academy of Sciences of the U.S.S.R. and president of the then powerful V.I. Lenin All-Union Academy of Agricultural Sciences) and his deputies wrought incalculable damage on biology and its application to agriculture. Brandishing self-righteousness, Lysenko and his followers fought with polemics and denunciations. Their attacks culminated in the removal of opponents from their positions (and usually much worse).

During this period Soviet agricultural programs were a constant embarrassment to the scientific community, and innovations and productivity lagged far behind other advanced nations. But Lysenko's views held sway and his power grew, even as Oswald Avery and his colleagues at the Rockefeller University were publishing evidence establishing not only the integrity of orthodox genetic theories but proof that DNA was the actual hereditary substance.[2] Only with the political demise of Khrushchev in 1964 were Lysenko's doctrines finally discredited. Even then, the scars of three decades of Soviet biology's pursuit of bizarre and aberrant science were slow to heal. Z.A. Medvedev observed five years after Lysenko's fall:

> Lysenkoism is far from having been liquidated; nor has it lost its aggressiveness. Neither has it lost from its midst people capable of grasping and comprehending modern biology, biochemistry and genetics, and capable of real education, yet unwilling to relinquish the primitive collection of dogmas they have so firmly mastered and held for so long. What is more to the point, they were also unwilling to relinquish the high posts they had occupied for so long (by no means because of their high qualifications).[3]

The Savaging of U.S. Biotechnology

It is almost inconceivable that in more open, democratic societies anything could emerge to rival the vicious and destructive power of Lysenkoism. But this does not mean that we should ignore significant parallels, or that Western democracies are immune to antiscience demagoguery.

Once again—this time in the United States—new science is being denounced to save old myth. In Lysenkoism, the old myth was an amalgam of Lamarckian theory and Communist dogma. In the Clinton administration—as applied to the new biotechnology, as well as to environmental issues generally—the myth is Greenspeak, the belief in a childlike, "natural" world of purity and innocence that is corrupted by scientific advances which "tamper with Nature." Beyond that the differences are more of degree than kind.

Vice President Al Gore, the point-man for the administration on technology policy, exhibits an ideological and exclusionary approach to policy making. At best, ideology makes scholarly neutrality difficult, if not impossible, for it serves a political function and not a truth-discovering function. With Gore in charge we are witnessing again an ideology imposed on science and single-mindedly driving public policy; a lack of understanding of science (specifically, of either its method or phenomenology); and an intolerance of dissenting views.

Al Gore: New Age Philosopher, Psychologist, Environmentalist

Gore's views inexorably shape the administration's policies on matters of technology (the president understands little of these matters, not surprisingly, and the weak presidential science advisor too defers routinely to the vice president). At least as applied to biotechnology these views are relentlessly negative. His endorsement of antibiotechnology diatribes and his criticism of biotech advances, as congressman, senator and vice president show a distinct lack of insight. More troubling is the fact that his writings reflect a general view that seems to place science and technology at odds with "the natural world" and by inference, the well-being and progress of humankind. Gore's book, *Earth in the Balance*,[4] provides a considerable and disturbing insight into the thinking of America's technology potentate.

Gore's apocalyptic central thesis is that we need to take "bold and unequivocal action...[to] make the rescue of the environment the central organizing principle for civilization." (starting at p. 269) Throughout the book, he uses the metaphor that those who believe in technological progress are as sinister, and polluters are as evil, as the perpetrators of the World War II Holocaust (see, for example, pp. 177, 196, 232-33, 256-57, 272-75, 281-82, 285, 294, 298, 366):

> It is not merely in the service of analogy that I have referred so often to the struggles against Nazi and communist totalitarianism, because I believe that the emerging effort to save the environment is a continuation of these struggles, a crucial new phase of the long battle for true freedom and human dignity (p. 275).

In *Earth in the Balance*, the vice president examines what he postulates are the political, eco-nomic (get it?), psychological, sociological and religious roots of the pollution problem. His economic section posits that "[c]lassical economics defines productivity narrowly and encourages us to equate gains in productivity with economic progress. But the Holy Grail of progress is so alluring that economists tend to overlook the bad side effects that often accompany improvements." (p. 188) This shortcoming of markets he considers "philosophically...similar in some ways to the moral blindness implicit in racism and anti-Semitism." (p. 189) Part of Gore's remedy is to redefine the relevant measures of a nation's economic activity. The purpose of this is made clear: it would enable the government to obscure the costs of environmental protection by calling them "benefits" and force businesses to list as societal "costs" some wealth-creating activity that would normally be considered beneficial to society (p. 343).

There is ample evidence that Gore has already begun to insinuate these ideas into both the Nation's foreign and domestic policies.

During a speech at Stanford University in April 1996, Secretary of State Warren Christopher announced a major new foreign policy initiative that will make environmental concerns coequal with American economic and national security in United States foreign relations. The initiative will, in effect, enlist the might of the entire United States' diplomatic apparatus to proselytize for vice president Gore's own ideological, idiosyncratic—and often scientifically insupportable—version of environmentalism.

A prime example is the Biodiversity Treaty (more formally, the Convention on Biological Diversity, or CBD, discussed in detail in chapter 4), which Christopher singled out as important and which Undersecretary of State Tim Wirth characterized as having "top priority among all treaties" and agreements awaiting confirmation. This international agreement is a volatile combination of poor-quality science and flawed environmental and foreign policy that, if implemented, would be bad for the United States and disastrous for the countries of the developing world. The new State Department initiative, as it applies to the CBD, would have the United States "fronting" for the United Nations and promoting it as the world's biosafety police force.

Backed by bloated and inefficient bureaucracies, and undeterred by their own meager scientific expertise, UN officials are now jostling to become international super-regulators. The immediate target of a half dozen UN programs or agencies is biotechnology, which they are eager to burden with a sweeping variety of new and unnecessary regulations.

At a time when hunger remains a serious problem for perhaps a billion people—the *real* endangered species—and per capita yields of the major cereals have leveled off or decreased, biotechnology holds great promise for raising productivity. The U.N.'s regulatory and policing proposals will nip this promise in the bud. The result will be UN-facilitated disease and famine. Among those threatened most by this regulatory impulse are the Asian countries that have made substantial investments in agricultural biotechnology. The list includes the Philippines, Thailand, India, Singapore and China, whose program is among the largest and most vigorous in the world.

One of the new biotechnology's great advantages is that its benefits are not limited to the industrialized world. Unlike many new technologies of our age, biotechnology has been available almost immediately outside the West. Since it builds on traditional agriculture and microbiology to help improve regionally important crops, biotechnology leads directly to that highest goal of developmental politics: self-sufficiency.

But a burdensome international bureaucracy enforcing ill-conceived regulation will stall, and even block, many of these benefits. Agricultural biotechnology is particularly vulnerable. It's an area where innovation is high but market incentives are often small and fragile. Vastly increased paperwork and costs for field testing will be potent disincentives for R&D in many countries.

The U.N.'s Trojan horse here is the CBD, discussed in detail in chapter 4. A product of the UN Conference on Environment and Development held in Rio de Janeiro in 1992, the CBD covers a wide range of issues related to the protection of biological diversity and the conservation of habitats in developing nations. At its conception, the CBD was advertised as an unprecedented opportunity to reconcile issues of conservation and access to biological resources. However, its good intentions are overshadowed by something called the international biosafety protocol. That's another name for biotechnology regulations, which do nothing to address threats to biodiversity. The real threats are from the introduction of "exotics," that is, organisms from other ecosystems; and from the expansion of farming to poorer-quality land.

Take field trials of improved varieties of potatoes, corn, rice or cassava, for example. No one anywhere would be allowed to grow and test a biotechnology-derived crop or garden plant—even on a plot of several square yards—without prior approval from regulators, on a case by case basis. Paperwork, red tape, politics and corruption would dog the process from beginning to end, from the first seed to the store shelves.

Ironically, many proposed UN biotechnology regulations will actually harm the environment. They will stifle the development of environment-friendly innovations that can help clean up toxic wastes, purify water and replace agricultural chemicals. Consider that plant genetics has already replaced some chemical pesticides by creating pest- and disease-resistant varieties of wheat, rice, soybeans, corn and other staple crops—but this process will be slowed or stopped by UN policies.

The State Department initiative will bring to the world in several new ways Vice President Gore's inimitable view of environmentalism. Selected U.S. embassies will establish "environmental hubs" to advance environmental goals, according to Christopher. Some see this as a valuable opportunity to make widely available to foreigners a vast array of ecological and environmental information—computerized data on and models of erosion, population growth, climatic patterns and so forth. But given the predilections described in this chapter, it is likely that the Clinton administration's State Department will be pushing Al Gore's version of such issues as ozone depletion, global warming, sustainable agriculture and biotechnology regulation. Whatever its virtues, the State Department is not charged to do a great deal of independent thinking; particularly on arcane scientific issues, it can hardly be expected to do more than support and promulgate the party line.

Even prior to the State Department initiative, Gore had already begun to insinuate these ideas into U.S. foreign policy. The U.S. Agency for International Development (USAID) has provided a kind of "slush fund" for the schemes of radical environmentalists. USAID foreign aid funds—derived, of course, from American taxpayers—are being used to undermine market economies abroad and put American businesses at a competitive disadvantage. In Indonesia, for example, USAID gave more than $1.3 million — virtually the entire operating budget—to the Wahana Lingkungan Hidup Indonesia (WALHI), the local chapter of Friends of the Earth. For the past 2 years, WALHI has campaigned against New Orleans-based Freeport McMoRan Copper & Gold, accusing the mining company of polluting an Indonesian river, destroying crops and inciting military attacks on civilians. But none of these accusations against the company has been substantiated. In addition, through U.S. environmental activists, WALHI successfully lobbied the Overseas Private Investment Corporation (OPIC), a federal agency that promotes business abroad by insuring companies against the risk of nationalization, to cancel Freeport's $100 million policy.[5]

In order to support and create the base of information necessary to justify his views, Gore has enlisted the resources of the Nation's intelligence community. In a speech at the World Affairs Council in Los Angeles on July 25, 1996, John Deutch, director of the CIA and the coordinator of all U.S. intelligence activities, signed on, "Just as Secretary Christopher promised 'to put environmental issues in the mainstream of American foreign policy,' I intend to make sure that Environmental Intelligence remains in the mainstream of U.S. intelligence activities. Even in times of declining budgets we will support policymakers..."

Gore has introduced his green agenda into other, domestic workings of the government. Consider this item on a White House Internet World Wide Web page called, "Al Gore and the Environment:"

> In April of 1994, the Department of Commerce Bureau of Economic Analysis (BEA) announced its economic-environmental accounting framework, or "Green GDP"—an effort initiated by the President in his Earth Day 1993 remarks. Natural resources and environmental quality are

> important productive assets that must be preserved in a healthy economy. Just as the economic ledger includes an entry for depreciation of plant and equipment, an entry should also be made to record the degradation of natural assets. Recording the full range of these costs, or the value of investing in environmental improvement, provides a clearer picture of the nation's wealth and a measure of sustainable income.[6]

According to this accounting, a paradigm of Orwellian doublespeak, the value of electricity produced by Hoover Dam could be considered a *cost* and a federal grant to Greenpeace as government *income.*

Gore's Green ideology can now be spread not only through our own government but throughout the world, courtesy of the official United States' diplomatic and intelligence apparatus. And this will happen in a way difficult for citizens—or even Congressional oversight committees—to monitor.

All environmentalism is not created equal. Al Gore's version is particularly fervent, paranoid and ill-informed. Gore has excoriated plant breeders' decision to develop herbicide-resistant plants that would decrease the overall use of herbicides and enable farmers to substitute less hazardous chemicals, saying that it "lent credibility to those who argued that biotechnology would make things worse before it made things better."[7] Gentleman farmer Thomas Jefferson knew better, when he said, "The greatest service which can be rendered any country is, to add a useful plant to its culture."[8]

Gore's New Age philosophizing leaves few clichés unexplored. Gore argues in *Earth in the Balance,* for example, that our approach to technological development has been shaped by aggressive male domination instead of by the nurturing instinct of women. "Ultimately, part of the solution for the environmental crisis may well lie in our ability to achieve a better balance between the sexes, leavening the dominant male perspective with a healthier respect for female ways of experiencing the world" (p. 213).

Gore's radical solutions to the costs and shortcomings of markets are government intervention and redefining words like "growth" and "productivity." Intrusion by the wise and enlightened government is necessary to save us from ourselves because

> ...our civilization is, in effect, addicted to the consumption of the earth itself. This addictive relationship distracts us from the pain of what we have lost: a direct experience of our connection to the vividness, vibrancy and aliveness of the rest of the natural world. The froth and frenzy of industrial civilization mask our deep loneliness for that communion with the world that can lift our spirits and fill our senses with the richness and immediacy of life itself (pp. 220-21).

Gore the psychologist continues:

> The unprecedented assault on the natural world by our global civilization is also extremely complex, and many of its causes are related specifically to the geographic and historical context of its many points of attack. But in psychological terms, our rapid and aggressive expansion into what remains of the wildness of the earth represents an effort to plunder from outside civilization what we cannot find inside. Our insatiable drive to rummage deep beneath the surface of the earth, remove all the coal, petroleum and other fossil fuels we can find, then burn them as quickly as they are found—in the process filling the atmosphere with carbon dioxide and other pollutants—is a willful expansion of our dysfunctional civilization into vulnerable parts of the natural world. And the destruction by industrial civilization of most of the rain forests and old-growth forests is a particularly frightening example of our aggressive expansion beyond proper boundaries, an insatiable drive to find outside solutions to problems arising from a dysfunctional pattern within (p. 234).

Gore the philosopher disparages the Cartesian method, the essence of scientific inquiry, for disconnecting man from nature. He also expresses disdain for Francis Bacon, who

> ...argued that not only were humans separate from nature; science, he said, could safely be regarded as separate from religion. In his view, "facts" derived through the scientific method had no moral significance in and of themselves [and]...the new power derived from scientific knowledge could be used to dominate nature with moral impunity (p. 252).

Gore concedes that this shift in thinking was essential for ending the Dark Ages and, eventually, for launching the Enlightenment. However, he thinks Plato and then Descartes and Bacon broke an important "spiritual triangulation" in human thought that identified the natural world as sacred, because, Gore believes, "each rock and tree was created by God." (p. 255) Gore argues that Bacon's "moral confusion...came from his assumption, echoing Plato, that human intellect could safely analyze and understand the natural world without reference to any moral principles defining our relationship and duties to both God and God's creation" (p. 256).

To this "error," which lies at the very heart of scientific objectivity, Gore attributes the responsibility for, among other evils, the atrocities of Hitler and Stalin:

> It is my view that the underlying moral schism that contributed to these extreme manifestations of evil has also conditioned our civilization to insulate its conscience from any responsibility for the collective endeavors that invisibly link millions of small, silent, banal acts and omissions together in a pattern of terrible cause and effect...*But for the separation of science and religion, we might not be pumping so much gaseous chemical waste into the atmosphere and threatening the destruction of the earth's climate balance* (emphasis added) (p. 257).

But for the separation of science and religion, we might still be burdened with the pre-Copernican notion that the sun and the planets revolve around the Earth.

The guiding principle of Gore's environmental agenda as set out in his book is, by any standard, radical: "We must make the rescue of the environment the central organizing principle for civilization."(p. 269) He goes on to draw analogies between the effort needed for this mission and the mobilization of this country to defeat the Nazis and to win the Cold War. He suggests an environmental "Marshall Plan" that might require $100 billion. (Consider, however, the adverse "income effect" on Americans' health and well-being of a government expenditure of this magnitude; see chapter 4.)

Gore is not so intoxicated with his own rhetoric that he believes his views are currently centrist. He believes that his views are morally in advance of those of the populace:

> [i]n any effort to conceive of a plan to heal the global environment, the essence of realism is recognizing that public attitudes are still changing—and that proposals which are today considered too bold to be politically feasible will soon be derided as woefully inadequate to the task at hand. Yet while public acceptance of the magnitude of the threat is indeed curving upward—and will eventually rise almost vertically as awareness of the awful truth suddenly makes the search for remedies an all-consuming passion—it is just as important to recognize that at the present time, we are still in a period when the curve is just starting to bend. Ironically, at this stage, the maximum that is politically feasible still falls short of the minimum that is truly effective...It seems to make sense, therefore, to put in place a policy framework that will be ready to accommodate the worldwide demands for action when the magnitude of the threat becomes clear. And it is also essential to offer strong measures that are politically feasible now—even before the expected large shift in public opinion about the global environment—and that can be quickly scaled up as awareness of the crisis grows and even stronger action becomes possible (pp. 304-05).

He assumes that he will initially have to settle for the partial measures that are politically possible, but that bold and dramatic measures are at the ready, awaiting the education and enlightenment of the public. This conviction, redolent of the "vanguard of the proletariat" working to raise the consciousness of man, is a stratagem that Marx, Engels and Lysenko would have understood.

The Assault on the New Biotechnology

On September 30, 1994, President Clinton formally handed Gore the leverage to make his regulatory *Weltanschauung* a reality when he signed an executive order allowing Gore, as the *Washington Post* reported, "to set the administration's agenda of regulatory priorities."

Biotechnology has been severely Gored by this administration. Like the policies of Lysenko, U.S. regulatory policies of the past three years, crafted with the full approval and collaboration of the administration (sometimes, indeed, at its direction), are based not on scientific consensus (in fact, quite the opposite) but on ideology that debases both scientific knowledge and common sense.

Regulatory agencies, by their very nature it seems, are always empire-building and in search of potential new responsibilities. What is new in this administration—and it is an exceedingly important change—is that there is no counterbalancing force for moderation, in the White House or elsewhere. There are no champions of either science-based approaches or of free-market policies occupying positions of power in the administration. Only the Republican-controlled 104th Congress's pressure for more rational regulation and reform of existing policies has averted a near-catastrophe for biotechnology.

Gregory Simon, Gore's domestic policy advisor, bragged in a speech that the administration had assembled "the best team you could assemble on biotechnology." The members he cited included himself, White House Science Advisor Jack Gibbons, FDA Commissioner David Kessler, and of course, team leader Al Gore. In the previous essay I discussed some of the flaws in this team. It is revealing that the administration should describe in such glowing terms a group that has collaborated on such a sorry (and preventable) outcome.

The Purges

As troubling as the substance of their policies is the mean-spirited nature of this administration's practices. The Clinton administration's regulatory and other policies—born of heterodox ideology akin to Lysenko's—have exerted a markedly negative impact on biotechnology research and development. And, as in the Lysenko regime, a good deal of the damage has been caused by self-righteous attitudes and vengeful actions. Gore et al have brooked no dissention or challenge to their view of policy or scientific rectitude. In fact, they have gone to extraordinary lengths to purge their "enemies."

In order to slant federal science and technology policy and to rid the civil service of dissenting views, the vice president and Greg Simon have interfered in federal personnel matters in ways that are, at the least, unethical. For example, while working for the vice president, Simon threatened a high ranking official at the Department of Energy with retaliation if she were to hire David

Kingsbury, the former assistant director of the National Science Foundation. (Simon and Kingsbury had clashed on biotechnology policy in earlier years. In fact, as a congressional staffer, Simon had hounded Kingsbury from government with unsubstantiated charges of conflict of interest.) Also while working for the vice president, Simon improperly ordered FDA to remove a senior civil servant at the Food and Drug Administration from his position. FDA officials admitted that this was retribution for the "transgression" of having implemented Reagan-Bush policies effectively. Gore himself dismissed Will Happer, a senior scientist at the Department of Energy, because he refused to ignore scientific evidence at hand that conflicted with the vice president's pet theories on ozone depletion and global warming. Similar incidents have been reported at the departments of State, Energy and Interior, and at EPA. In these departments and agencies a number of prominent civil servants have been moved to less visible positions or substituted with other officials during interactions with the White House for their own "protection."[9]

The low road that Gore takes to science and its role in public policy serves as a reminder that ignorance is not a simple lack of information but an active aversion to knowledge—the refusal to know—issuing from hubris or laziness of mind. More disturbing still, because of its practical implications, is his philosophy of government, particularly with respect to the federal oversight of new technology and environmental protection. Gore's views are (to borrow a phrase from George Will) paradigmatic of paternalistic liberalism, of government that is bullying because it is arrogant, and arrogant because it does not know what it does not know. The president promised "leaner but not meaner government," but what we've gotten is quite the opposite.

In his discussion of a series of eccentrics in 20th century Soviet science, Robert Conquest observed:

> Not all Marxists would have been capable of such powerful idiocy. But to some degree Marxism itself can certainly be blamed. First of all, the principle of accommodating science to a particular metaphysic rather than leaving it to act autonomously seems bound to produce distortion. Secondly, the notion that Marxism is a basic universal science leads to the condition in which many people

> professing it feel that they are already fully educated and, in effect, capable of judging any subsidiary studies without adequate humility or effort. Hence, perhaps, its attraction for a certain type of Westerner.[10]

More important, perhaps these notions tend to show that those who seek cure-all formulae for reconstructing society are temperamentally inclined towards 'unorthodox' fads in other fields. This in turn may tend to cast doubt on the validity of their political-economic analyses.

These same observations are applicable to Gore's thinking and approach to policymaking, where environmentalism is the equivalent of Marxism. Gore is certainly culpable of attempting to distort science to fit a particular metaphysic, of know-it-all arrogance, and of seeking cure-all formulae for reconstructing society. He is, therefore, vulnerable to the same pitfalls as his wrong-headed forebears, and both citizens and commerce in this country have already begun to bear the consequences of his powerful idiocy.

Mr. Gore must be made to realize that science is not politics. Politics is the art of the possible, while science is the search for truth.

CONCLUSIONS

Vice President Gore's view of science is sadly reminiscent of the academicians of the imaginary city of Lagado in Jonathan's Swift superb parody, *Gulliver's Travels.*[11] Gore would likely have welcomed a "photo-op" with the researcher who "had been Eight Years upon a Project for extracting Sun-Beams out of Cucumbers, which were to be put into Vials hermetically sealed, and let out to warm the Air in raw inclement Summers." This is precisely the kind of "sustainable agriculture" that Gore's constituency would wish the government to subsidize, in lieu of genetic engineering.

But the most apt metaphor in *Gulliver's Travels* for Gore's view of science is an "operation to reduce human Excrement to its original Food, by separating the several Parts, removing the Tincture which it receives from the Gall, making the Odour exhale, and scumming off the Saliva." Here we have the kind of recycling technology that is the stuff of Clinton administration executive orders (or should that be *ordures*?)—literally, the stench of flawed science and bad public policy in the nostrils of reasonable people.

While the similarities can be overdrawn, to be sure, there are disturbing parallels between the Lysenko era of Soviet genetics and current biotechnology regulatory policy in the United States. Once again, this time in the United States, new science is being denounced to save old myth. In Lysenkoism, the old myth was an amalgam of Lamarckian theory and Communist dogma. In the Clinton administration, as applied to the new biotechnology, the myth is a pristine and pure natural world despoiled by science, technology and industrial development.

The Clinton administration's approach to regulating the purported risks of the new biotechnology is inconsistent with—indeed, is diametrically opposite to—the worldwide scientific consensus that biotechnology oversight should be consistent with scientific principles and risk-based. In a relatively short time, Vice President Gore's influence, in particular, has damaged whole sectors of research and commercial application of the technology. Bioremediation, agricultural biotechnology, food production, and iopharmaceuticals have all suffered because of regulatory or other disincentives. Not even biotechnology's most environment-friendly manifestations, such as biocontrol agents and new plant varieties that could supplant agricultural chemicals, have been spared.

Redolent of Lysenkoism are Clinton administration officials' misguided certitude, mean-spiritedness, intolerance to dissent on scientific issues and hostility to pluralism in policy making. But while Lysenko was ill-educated and charismatic, Gore is highly educated (some would say, overeducated) and deadly dull.

By means of appointments to federal advisory committees, the hiring and promotion of certain tenured civil servants, the setting of funding priorities, the profusion of propaganda from federal agencies and formal rulemaking that establishes policies and regulations, the Clinton administration will exert influence beyond its tenure.

Will the administration's mischief prove to be so pervasive and injurious that we have to wait decades for its effects to run their course, as in the Soviet Union? U.S. industry is currently highly competitive in most areas of biotechnology (thanks partly to stultifying regulation in Europe and Japan), but, in the wake of the missteps of Gore and his allies, that could change quickly.

How much will Americans lose in jobs and new consumer products? Along with threats to U.S. competitiveness, illogical and burdensome regulatory policies will bring higher prices, greater reliance on chemicals for agriculture and bioremediation, fewer benefits and choices for American consumers—and even more dire straits for researchers. And what of the intangible effects of government policies that exhibit contempt toward rationality and science?

Decisions pertaining to science, technology and risk should, insofar as it is possible, be independent of political ideology. But as we shall see in subsequent chapters, the Clinton administration's highly ideological approach, reminiscent of the Lysenko excesses, moves us far from that ideal. It places citizens' interests increasingly at the mercy of demagogues and bizarre political compromises. Historian Barbara Tuchman observed that, "Mankind... makes a poorer performance of government than of almost any other human activity."[12] Unhappily, Lysenkoism, American-style, provides supporting evidence for her assertion.

References

Essay 1: Introduction

1. Anon. Introduction of Recombinant DNA-Engineered Organisms into the Environment: Key Issues. Washington D.C.: National Academy Press, 1987.
2. Anon. Field Testing Genetically Modified Organisms: Framework for Decisions. Washington D.C.: National Academy Press, 1989.
3. Hofstadter R. The Paranoid Style In American Politics. Chicago: Univ. Of Chicago Press, 1952.
4. Huttner, SL. Government, researchers and activists: the crucial public policy interface. In: Brauer D, ed. Biotechnology. Weinheim, Germany: VCH, 1995:459-494.
5. Meister, I. Uncontrolled Trade in Genetically Manipulated Products. Press release, Greenpeace International. April 7, 1995.
6. Idem.
7. Kennedy D. The Regulation of Science: How Much Can We Afford? MBL Science, Winter 1988-89:5-9.
8. Idem.

Essay 2: You Call *This* a Pro-Technology Policy?

1. Anon. One Tough Government Downsizer. The American Enterprise 1996; 11.
2. Clinton WJ, Gore A. Science in the National Interest. Washington D.C.: Executive Office of the President, Office of Science and Technology Policy, 1994.

3. Haas E. Diet Risk Communication: A Consumer Advocate Perspective. In: Gaull GE and Goldberg RA, eds. The Emerging Global Flood System. New York: Wiley & Sons, 1993:133-46.
4. Idem.
5. U.S. Congress, Office of Technology Assessment. New Developments in Biotechnology—Field-Testing Engineered Organisms: Genetic and Ecological Issues, OTA-BA-350. Washington D.C.: U.S. Government Printing Office, May 1988.
6. Clinton HR. It Takes a Village. New York: Simon & Schuster, 1995.
7. Davis B. Personal communication. Also, Hoyle R. Comments from the White House's Greg Simon. Bio/Technology 1993; 11:1504-5.
8. Anon. National Biotechnology Policy Board Report. Bethesda, MD: National Institutes of Health, Office of the Director, 1992.
9. Gore A. "Jacket quotes" for Rifkin J. Algeny. New York: Penguin, 1983.
10. Gore A. Planning a New Biotechnology Policy. Harvard Journal of Law and Technology 1991; 5:19-30.
11. Idem.
12. Idem.
13. Anon. National Research Council News Report. February 1987; 37:1.
14. McNicoll G. The United Nations' long-range population projections. Pop Devel Rev 1992; 18:333-340.

Essay 3: Powerful Idiocy: Lysenko, Gore and U.S. Biotechnology Policy

1. Conquest R.. We & They. London: Temple Smith, 1980:78.
2. Davis BD. Molecular genetics, microbiology and prehistory. BioEssays 1988; 9:129.
3. Medvedev ZA. The Rise and Fall of T.D. Lysenko. New York: Columbia University Press, 1969:240.
4. Gore A. Earth in the Balance. New York: Plume 1992; page references will be given parenthetically in the text.
5. McMenamin B. Environmental Imperialism. Forbes May 20, 1996.
6. World Wide Web page, "Al Gore and the Environment:" http:/www.whitehouse/gov/wh/eop/ovp/html/Enviro@GDP.html, May 1, 1996.
7. Gore A. Planning a New Biotechnology Policy. Harvard Journal of Law and Technology 1991; 5:24-25.
8. Jefferson T. The Papers of Thomas Jefferson. (Accession No. 39161) Library of Congress, 1800.
9. See Miller HI. Gore and His Minions Punish Civil Servants Who Dare to Disagree. Washington Times, June 2 1994. Also Davis B, personal communication.
10. Conquest R. Ibid., 160.
11. Swift J. Gulliver's Travels. New York: Bantam, 1962:177-89.
12. Tuchman B. The March of Folly: From Troy to Vietnam. Boston: Knopf, 4.

CHAPTER 3

THE VAGARIES OF FEDERAL REGULATION

"Washington's motto: 'Don't laugh, you're paying for it.'"
—Humorist Dave Barry

INTRODUCTION

In the United States, government regulation is not organized around technological processes; rather it is structured around products and their intended uses. As uses vary, so do agency jurisdictions and evaluation procedures. Agency approaches are driven by several factors, including enabling statutes, regulations that implement the statutes, and case law and other precedents. As discussed in chapter 1, the diversity of biotechnology products and their uses triggers regulation by several agencies and evaluation procedures that vary widely in focus, nature and rigor. This is described in some detail in the federal government's "coordinated framework" for biotechnology regulation.[1]

FDA regulates the testing and marketing of biotechnology products under the Public Health Service Act and Food, Drug and Cosmetic Act. The agency regulates products that fit the definitions of drugs, medical devices (diagnostic test kits, for example) and food additives (such as sweeteners and preservatives). These products are subject to *premarketing* approval: the agency must make an affirmative decision before the product can be sold. Other products, such as foods themselves, are also regulated by FDA but are not subject to premarket review. Nonetheless, they may be removed from commerce or seized by FDA if they are found to be "adulterated" (that is, containing any addition "which may render

[them] injurious to health").[2] This is a form of *postmarketing* regulation.

The EPA regulates biotechnology products primarily under two statues: the Federal Insecticide, Fungicide and Rodenticide Act (FIFRA), which encompasses pesticides and growth modulators; and the Toxic Substances Control Act (TSCA), a regulatory catch-all that provides oversight of "nonnatural" and "new" substances or mixtures of substances intended for use in commerce and not regulated elsewhere. Both of these statutes require premarketing regulation.[3]

USDA regulates under separate statutes on the basis of either end-use or risk. Under the Virus, Serum and Toxin Act, the USDA regulates according to product end use—for example, veterinary vaccines. Under the Federal Plant Pest Act (FPPA), USDA regulates according to product risk—organisms that may be plant pests. The FPPA regulations contain lists of organisms known to pose a plant pest risk. The testing, transport or sale of any organism on these lists requires permission from the USDA.[4]

In this chapter it will become evident that federal regulatory agencies have jumped on the biotechnology bandwagon with different and sometimes incompatible, if not downright contradictory regulatory approaches. The only consistent feature of policies proposed since 1985 is that (with the partial exception of policies at the FDA) they selectively disadvantage biotechnology products in the competitive marketplace.

In some cases, the agencies have crafted scientifically dubious definitions of the scope of their regulations in order to create legal mechanisms for regulating the broadest possible array of new biotechnology products. For example, EPA's regulations under FIFRA and TSCA, and USDA's regulation under FPPA are all built on definitions that equate novelty with risk in inventive oversight schemes that capture virtually all plant and microbial biotechnology products intended for agricultural or environmental remediation purposes (*vide infra,* this chapter).

Agencies have often heralded inconsequential refinements in their regulatory systems while overlooking real opportunities for improvement. Some have claimed improvements when they have actually slowed down their regulatory processes. FDA has flip-flopped on its regulatory approaches for new biotechnology foods and dropped the ball on substantive drug regulation reforms demanded by Congress and by the Bush administration. USDA proposed a workable and laudable streamlined notification process for regulating rDNA modified plants and then simply failed to implement it.

I hope that the following essays will drive home for the reader the magnitude and kinds of impacts that government decisions and actions exert on industry and the public. In order to illumine certain vagaries of federal regulation, some of these include pertinent nonbiotechnology examples.

ESSAY 1
DR. KESSLER'S FDA: BUREAUCRATS' OWN INTERESTS COME FIRST

The costs of federal regulation are monumental. The General Accounting Office (GAO) has estimated the annual direct and indirect costs at approximately $647 billion, and combining the estimates of various scholars yields a similar result: a range of some $635-857 billion, or between $6565 and $8869 per household, annually.[1] Many Americans and their institutions have begun to wonder whether we get our money's worth. They have good reason to wonder. FDA, alone, regulates products accounting for 25 cents of every consumer dollar, including virtually all of biotech's high-ticket items.

With great fanfare in January 1995, FDA officials released their 1994 drug approval statistics. They proclaimed dramatic improvements in approval times for biologicals (a class of drugs that includes vaccines and biotechnology and blood products; discussed further, below). They also claimed that 15 "major biologics [approval] actions" had been taken. On closer inspection, it was much ado about nothing.

FDA's numbers game illustrates two practices at which regulatory agencies excel: "gaming" a situation and manipulating statistics. In so doing, they mask a different reality. It is akin to a quarterback enhancing his statistics on completions by throwing short passes that lose yardage.

In gaming the situation, FDA calculated drug approval times in a manner that obscures how minimal the improvement really is. Manipulating statistics, FDA reported medians instead of means (averages), thus minimizing the effect of statistical "outliers." Then, by a sleight-of-hand, FDA shortened the apparent evaluation time by beginning to report the time between receipt of the dossier and issuance of an "approvable" letter, rather than the critical "approval" letter which routinely comes months later. Because marketing of a product cannot begin until receipt of the *approval* letter, the FDA statistics tell a deceptive story of how long companies stay engaged in the regulatory process and deprived of entry into the marketplace. The FDA's unpublicized changes in how it reports its performance seem to provide no advantage—except to make the agency appear to be more efficient.

Consider how FDA's claim of 15 major biologicals actions in 1994 also stretches the truth. About half were anything but "major." Three involved diagnostic kits, two involved a change in dosage for a previously-approved drug, one was a blood bank reagent, and one was a detergent inactivation process for viruses. Only four products—two vaccines, a cardiac drug and a cancer therapeutic—could be considered of even moderate importance, and the vaccines are similar to others already on the market. This was the *entire* annual output (not just biotechnology products) of one of the principal regulatory units of FDA, an agency with a billion-dollar-plus budget.

Approvals of biotechnology-derived biologicals have been especially disappointing. Only one was approved in 1994 and none in 1995. An appreciation of how truly dismal the agency's overall performance has been requires a little background on the drug approval process.

The Slowing Pipeline of Drug Approvals

The sequential approval process begins with FDA evaluation and approval of the first clinical trials of a product in normal volunteers or patients. This approval initiates the second step, the clinical testing process (under FDA oversight) which can require a decade and cost hundreds of millions of dollars.[2] An important juncture is reached when the manufacturer deems that the testing has demonstrated safety and efficacy for a certain disease and submits an application to FDA for marketing approval, initiating the third step. FDA's evaluation for the final approval may require a couple of years or more. Recently this last phase has been progressing somewhat faster. Overall, however, the process is lengthy and slow, and the absolute number of products approved has been flat for several years.

Rather than improving the process overall, FDA has systemically introduced various new obstacles to pharmaceutical development during the Clinton administration. Concomitantly, FDA has reported a significant decline in the numbers of new applications for marketing approval of biologicals (which includes most new biotechnology drugs).

How, exactly, has the FDA slowed movement through the pipeline? The accumulated policy changes and product-related decisions have proven to be the regulatory equivalent of "death by a thousand cuts" for drug companies. The agency has introduced extraordinarily stringent rules on the reporting of drug side effects, on manufacturing, and on the "promotional activities" by which companies disseminate new research results to physicians. The changes in manufacturing rules are of particular concern to companies because they can cause significant interruptions in drug production during clinical trials for even the most trivial change in procedure. All of the changes stem from a hyperregulatory, zero risk FDA mindset that inexorably drives upward the stringency, expense and intrusiveness of drug regulation.

Relying on an increasingly active and sophisticated propaganda apparatus (the Office of Public Affairs, headed by a Gore loyalist), FDA has put a positive spin on this eroding situation—reporting that the approval process has sped up, that FDA has solved its problems, that critics of the agency are behind the times. The truth is that with fewer incoming applications, FDA reviewers had more time to spend on each, and the final approval step proceeded in a more timely manner. FDA's method of reducing its backlog is like a District Attorney who requests dismissals for all pending felony indictments and then claims credit for reducing his caseload.

There is little to celebrate in this. There are fewer applications because FDA's new regulatory barriers have stalled drug manufacturers in earlier stages of development and evaluation. The apparent improvements in final approval times are simply an artifact of the regulatory burdens the agency recently implemented. With the help of the Clinton administration, FDA has improved the numbers at the cost of product innovation, business development and patients' well-being.

Is it any wonder, then, that the process of drug development in the United States has become the longest and most expensive in the world? From synthesis of the therapeutic molecule to marketing approval, the entire process requires 12-15 years, at a cost averaging half a billion dollars.

FDA's shortcomings are largely attributable to poor management superimposed on a system plagued by distorted incentives, disincentives, rewards and punishments for agency officials. Part of this is the compulsion of regulatory reviewers to avoid errors of commission ("Type 1" errors, discussed in chapter 5) at any cost. Under David Kessler's leadership, the stultifying overregulation of foods, drugs and medical devices at FDA has become acute.

When he became commissioner in 1991, Kessler found the agency to be ponderous and slow. He promised dramatic changes, that he "would teach the elephant to dance." And Kessler has—to a tune called by flacks and politicians. Right out of the gate, Kessler chose a high-profile but ludicrous case that was calculated to get him on the evening news in a virile demonstration that he is tough on industry. Did he choose defective heart valves, a contaminated vaccine, a drug causing sudden death? No, Citrus Hill orange juice. Kessler commanded federal marshals to confiscate 15,000 gallons of juice. Was it spoiled, contaminated, unfit for human consumption? Nope. It was labeled "fresh" when it was actually made from concentrated orange juice. Federal guidelines say it is inaccurate to call orange juice "fresh" if it's made from concentrate.

On CBS' *60 Minutes*, Kessler expressed his indignation: "[the juice] was made from concentrate. My grandmother could have told you, I mean, it wasn't fresh. It wasn't very hard [to tell the difference]." Aren't consumers justified in asking Dr. Kessler: "If we can so easily tell by taste that the juice came from concentrate, why not simply let us decide whether we like the product well enough to buy it again? Why aren't you more concerned about our taxpayer dollars footing the bill for FDA's regulatory compliance staff and lawyers, and for the federal marshals who corralled the outlaw juice?"

One suspects that Kessler *grand-mère* would have had more sense.

But this is "get tough" policy, Kessler style. He admonished his staff and said publicly that FDA is an "enforcement agency" and that, by God, industry will know it. (Meeting privately with industry, however, Kessler took a different line. He explained that the bluster about enforcement was just for show, to keep FDA's left-wing critics off his back. He assured them that he was very sympathetic to industry.)

Unfortunately, as we have seen in the case of drug regulation, Kessler has gone beyond the posturing of the silly Citrus Hill affair. He personally master-minded new impediments to drug development. For certain particularly ill-advised and regressive regulatory proposals (such as the food biotech requirements and the new requirements for reporting of adverse reactions, both discussed below), Kessler was almost the only booster within FDA.

The added costs, delays and outright barriers have drawn a chorus of protests and elicited serious demands for regulatory reform. The roster includes even some of the usual defenders of big government and heavy-handed regulation, such as Congressman (now Senator) Ron Wyden (D-OR), a member of the House Commerce Committee, who proposed a list of substantive reforms; and Senator Barbara Mikulski (D-MD), who administered a tongue-lashing to Kessler during a 1995 Senate Labor and Human Resources Committee hearing.

Facing the prospect of Congress reforming his agency, Kessler has been running scared. In March 1994, President Clinton promised changes. Kessler did an abrupt and very public about-face, claiming to be a born-again regulatory reformer. Together, Kessler and the Clinton administration have used every opportunity to demonstrate that conversion. Nonetheless, the promised reforms have ranged from the modest to the insubstantial. It is not unduly cynical to predict that even these small reforms will evaporate if the 1996 elections enable the Democrats to control both the White House and at least one house of Congress.

Pseudoreforms Ignore Restrictive Policies

It is instructive to consider what the Clinton administration has *not* included in the proposed reforms. They have chosen not to roll back regressive policies that they only recently implemented. They have left in place broad new guidelines requiring certain numbers of women and minorities in federally funded clinical research—a requirement that will dramatically slow and increase the cost of clinical trials, and encourage companies to do clinical research abroad.[3] (For technical reasons, this requirement was actually promulgated by NIH, not FDA.) They have not withdrawn burdensome new proposed rules on the reporting of drug side effects.[4] They have left intact recent restrictions on "promotional activities."[5] The latter two policies are sufficiently contrary to the public interest that they deserve amplification.

Regulations proposed by FDA in October 1994 would require clinical researchers to report drug side effects more quickly and to presume that the new drug is to blame any time a patient gets sick or sicker during a study.[6] The regulations reconfigure the burden of proof so that all new drugs are construed guilty until conclusively proven innocent. They set a lower regulatory threshold for stopping a clinical trial, making the entire drug development process unnecessarily risk averse, slower and more expensive.

Comments on this proposal from academia and industry are instructive. The director of the Johns Hopkins University Center for Clinical Trials, Dr. Curtis Meinert, observed that the proposed regulations would "make individual IRBs [Institutional Review Boards] into data and safety monitoring committees," a role for which they are ill-equipped. From a careful comparison of two clinical trials, one performed with reporting according to the old requirements and one under the new rules, he and his colleagues concluded that the FDA's proposed changes would increase the cost per patient and the paperwork generated per patient 62-fold and 63-fold, respectively. The Johns Hopkins study concluded that these costs were not accompanied by a commensurate benefit to patients.[7]

The DuPont-Merck Pharmaceutical Company estimated that, under the proposed regulations, its reporting burden would double for each prospective drug in development. Amgen, Inc. described the practical difficulties of estimating the expected incidence of death and serious adverse events that arise, not from the drug, but from underlying disease or concomitant medications.

It is noteworthy that this more burdensome regulatory process favors larger, established companies that have products on the market and that have a relatively broad resource base. Small, entrepreneurial companies (which make up the majority of biotechnology firms) typically have products in early developmental stages, few on the market and limited financial resources. In the highly competitive world of drug commercialization, FDA's proposed scenario disadvantages those with less experience and resources devoted to regulatory affairs.

Consider also the agency's more stringent approach to "promotional activities" in the context of the drug approval process. When the FDA grants approval of a drug, the agency specifies the disease, condition or symptom for which that approval is valid. That information—the "on label" use—represents the limit of the manufacturer's claims in labeling, advertising and promotion. However, approved uses are very narrowly defined and often much more restrictive than the data and clinical experience support. Based on their own experience and knowledge of the medical literature, physicians commonly prescribe broader, additional "off label" uses for diseases or symptoms other than those FDA has formally approved. (FDA does not prohibit this.)

Some 40-50% of all drugs are prescribed for "off-label" uses; that is, for diseases or symptoms other than those the FDA has approved. Sixty to 70% of drugs used to treat cancer and 90% of drugs used in pediatrics are for off-label uses.[8] For example, while the drug Taxol was long approved only for ovarian cancer, cancer specialists knew from published studies that it was also effective in breast cancer. However, the FDA prohibits not only the inclusion of such information in the drug's labeling and advertising, but even the dissemination of articles from medical journals that describe off-label uses.

Consider another similar example of the FDA's oversight of what it considers to be drug "promotion." It would not be unusual, of course, if a drug approved for short-term treatment of a disease were also beneficial for long-term treatment of the same disease. Physicians can freely share and use that fact, but if a drug company communicates the same information it is subject to FDA sanctions.

The last example is not hypothetical. An approved use for the drug Axid, produced by Eli Lilly and Company, is short-term (up to 12 weeks) treatment of gastroesophageal reflux disease, a particularly unpleasant condition marked by heartburn, belching, and often, difficulty in swallowing. In the process of gaining FDA approval for the short-term therapy, the company performed studies and submitted the resulting data both to FDA reviewers (who judged it to be adequate) and to the medical journal, *Digestive Diseases and Science*, which subsequently published it. In the journal article, the authors briefly described data indicating that long-term therapy at full dosage is frequently required to prevent relapse. When the company sent out copies of the journal article to physicians, FDA officials objected. The agency prohibited the company from disseminating the article even though the findings reported had been peer reviewed by medical experts before the journal accepted it for publication. FDA said that information on short-term use of the drug could be discussed because it is an approved on label use, but long-term treatment could not be discussed because it is off-label.[9]

FDA routinely prevents drug companies from informing doctors about almost half the uses of the drugs at their disposal, and almost all their uses in pediatrics.[10] The agency has become a potent censor of medical information that is often vital to patient care. Since 1962, FDA has interpreted in progressively broader and more divergent terms the statutory authority that enables it to regulate "drug promotion and advertising." The agency has increasingly arrogated power over how and what drug companies may communicate to health professionals and the public.[11] FDA Commissioner David Kessler has further extended that authority in extraordinary ways. FDA now interferes with the ability of physicians to present new findings at medical conferences and with companies' efforts to convene consumer focus groups. Focus groups help drug manufacturers obtain information on how to make product packaging and labeling more user friendly once new drugs are approved.

The impact of FDA's censorship and micromanagement is selective and socially regressive. Those health professionals who keep up with new developments in medical journals more commonly work in academic teaching hospitals as specialist physicians and medical students. They are expected to follow advances in patient care. It is noteworthy that teaching hospitals disproportionately serve middle and upper class patients.

In contrast, nonteaching institutions such as private hospitals or clinics provide medical care to a broader spectrum of patients and largely through general practitioners or nonphysician health professionals, such as nurses or physicians' assistants. These professionals are less likely to keep up with the specialized medical literature. They depend upon information conveyed through other professional or more informal channels. The barriers created by FDA censorship have disproportionate impact in this health care setting compared to the teaching hospital. They also, therefore, have disproportionate impact on patients from lower socioeconomic backgrounds. Those patients who can afford specialists will continue to have access to the latest advances in drug therapies, while those patients who rely on nonteaching hospitals and care from generalists will not, thanks to the FDA.

Why does FDA do it? It's in their self-interest (see chapter 5). Economist Milton Friedman has pointed out that the potent, ubiquitous force of self-interest is served by different actions in the private sphere than in the public realm:

> The bottom line is different. An enterprise started by a group of people in the private sphere may succeed or fail. Most new enterprises fail...If the enterprise fails, it loses money. The people who own it have a clear bottom line. To keep it going, they have to dig into their own pockets. They are reluctant to do that, so they have a strong incentive either to make the enterprise work, or shut it down.
>
> Suppose the same group of people start the same enterprise in the government sector and the initial results are the same. It is a failure; it does not work. They have a very different bottom line. Nobody likes to admit that he has made a mistake, and they do not have to. They can argue that the enterprise initially failed only because it was not pursued on a large enough scale. With the best intentions in the world, they can try to persuade the people who hold the purse strings to finance the enterprise on a larger scale, to dig deeper into the pockets of the taxpayers to keep the enterprise going. That illustrates a general rule: If a private enterprise is a failure, it closes down—unless it can get a government subsidy to keep it going; if a government enterprise fails, it is expanded.[12]

It's also in the nature of government officials to arrogate more and more functions—former FDA Commissioner Frank E. Young used to quip, "Dogs bark, cows moo and regulators regulate."

Overall, the Clinton administration's proposed changes seem mainly intended to impress the uninitiated and mollify some of the critics of FDA's overregulation, rather than to achieve meaningful reform. Echoing a point made earlier, a significant flaw in the "reforms" is that their meager benefits will be directed principally to larger, established companies that already have products on the market. Smaller, highly entrepreneurial companies (many of which are found in the biotechnology industry) whose products are largely in early developmental stages come up empty.

But perhaps the most serious flaw in the administration's plan—one that could vitiate impact entirely—is that the reforms are to be implemented *by FDA*. Experience should have taught the futility of an agency being directed to reform itself—especially with David Kessler at the helm. Similar reforms, instigated by President Bush's Council on Competitiveness, were announced by the FDA (near the end of Kessler's first year heading the agency) in 1991. To no one's surprise, the agency studied them, literally, to death. As an FDA official at the time, I recall agency officials' amusement at the prospect of reforming themselves—and that was during a presidential administration that really did care about streamlining regulation.

Miracles or Magic Show?

The FDA demonstrated particular bureaucratic cunning in a plan announced in November 1995, ostensibly to modernize and streamline regulation of a class of therapeutic products called "biological" drugs, or biologicals.[13] This was one of a series of reform "miracles" that FDA intended to advertise FDA's new commitment to reform. In truth, it's smoke and mirrors. A little background is necessary to understand why.

Biologicals (blood and blood products, vaccines, extracts of natural substances for treating allergies, extracts of living cells, and the like) are a subset of drugs. Historically, they have been regulated differently from other drugs. Biologicals used to be commonly derived from biological sources, including animals and microorganisms. They were often impure and their active substances usually poorly characterized, making it difficult to demonstrate that each batch meets specified standards of purity or potency. FDA has therefore required that biologicals producers demonstrate a high level of control over the manufacturing process. Every procedure in the production pathway has to be performed in a highly regimented and reproducible way.

FDA ultimately grants marketing approval based both on the product's demonstrated safety and efficacy and on certification of the manufacturing establishment's production control and rigor.

In addition, samples from every production batch must be submitted for certification by the FDA, which analyzes them for purity and potency (to the extent that such measurements are possible).[15]

In contrast to biologicals, other drugs are generally smaller and simpler molecules that are (most often) chemically synthesized, highly purified and well characterized. FDA regulates these drugs somewhat differently. Unlike biologicals, each batch of these drugs must meet certain analytical standards set by the agency. Neither licensing of the manufacturing facility nor batch certification is required.

Advances in technology have blurred the distinction between biological and other drugs, however. Many biologicals—particularly those made with the techniques of the new biotechnology ("gene-splicing," or recombinant DNA technology)—are now often highly purified, well-characterized preparations that can be regulated the same as other drugs. Frequently, the FDA has done just that—established strict standards which each batch must meet, exempted the products from certification of each batch, and so forth. Biologicals' manufacturing facilities continue to be certified, but in practice this is not very different from the inspections that are performed in the plants that make other drugs.

In November 1995 FDA codified this practice when it announced that henceforth it would no longer treat as "biologicals" those products that fall into the category of "well characterized biotechnology products." Thus, Kessler and the Clinton administration dressed up as new regulatory reform what had become near-standard practice. There is something rather ironic about FDA's making a fuss about beginning to regulate biotech-derived biologics like chemically-synthesized drugs: for most of the last 20 years, given a choice of whether a specific new biotechnology product would be regulated as a drug or as a biological, pharmaceutical manufacturers would have preferred the latter. They have considered the overall treatment afforded to sponsors by biologicals' regulators at FDA to be more collaborative, predictable, scientific and expeditious.

In a "bury the bone, dig up the bone" exercise in December 1995, the FDA presided over three days of dialogue with industry about the definition of a "well characterized biotechnology product."[15] This was another example of a high-profile, low-benefit FDA scheme—the moral equivalent of a politician's "photo-op." Leaving aside the question of FDA's sincerity, any benefits to biotechnology companies would be, at best, minimal: there have been only about 20 biotechnology-derived biologicals approved by the FDA during the past decade, and, as mentioned above, only one was approved in 1994 and none in 1995. A genuine and useful FDA reform of this kind would be to eliminate entirely the distinctions between drugs and biologics (perhaps with the exception of blood and similar products), and to modernize regulatory requirements for the consolidated group of products.

No matter how you look at them, the Clinton administration's "new" FDA policies, announced with great fanfare by no lesser personages than President Clinton, Vice President Gore, HHS Secretary Shalala and Commissioner Kessler, have largely been a hoax. One of the most highly touted—the "March (1996) miracle" that would "speed up the entire process" of cancer-drug development[16]—arrived the very day after three serious FDA reform bills were introduced in the House of Representatives. In it, FDA proclaimed that "[i]t is appropriate to utilize objective evidence of tumor shrinkage as a basis for approval, allowing additional evidence of increased survival and/or improved quality of life associated with that therapy to be demonstrated later." In other words, the agency would permit the use of surrogate endpoints for cancer studies.

Once again, there is little new here. The FDA long has had great flexibility in standards for overseeing drug testing and approval, including the use of "surrogate endpoints" for disease regression or prevention. Sometimes, improvement in a laboratory value is the major criterion used in judging a drug's efficacy. For drugs that lower serum cholesterol or blood pressure, for example, FDA no longer requires a demonstration that treatment actually increases survival or reduces the incidence of heart attacks and stroke. Significant improvement of "the numbers" is sufficient. Back in 1991, prodded by the Bush administration's Council on Competitiveness, FDA announced a policy of using "flexibility in the current statute to develop and adopt surrogate endpoints whenever possible to measure the efficacy of drugs used to treat life-threatening diseases." The announcement went on to assure that FDA would accelerate approval of drugs via

> the use of surrogate endpoints or other appropriate indicators of effectiveness to establish a drug's efficacy provided that the endpoint is a reasonable predictor of a clinical benefit. For example, under this reform, FDA will rely upon indications such as tumor shrinkage to show that a drug may treat cancer, rather than waiting for conclusive tests that patients live longer.[17]

A much stronger provision than FDA's current proposal was offered by HR 3199, the Drugs and Biological Products Reform Act introduced in March 1995. (As this manuscript went to press, the bill appeared to have been stonewalled and killed by the Clinton administration.) Echoing the promises of the FDA's 1991 announcement, for *all* "serious and life threatening" conditions HR 3199 would establish in law a new standard for marketing approval: the conclusion by experts that "there is a reasonable likelihood that the drug will be effective in a significant number of patients and that the risk from the drug is no greater than the risk from the condition." Most important, the extension of the new efficacy standard to diseases other than cancer is only humane: Why should patients with stroke, multiple sclerosis, Alzheimer's disease, emphysema, crippling arthritis or heart failure be deprived of benefit from this more rational standard?

The bill's new standard for drugs used in serious and life-threatening conditions is derived from the FDA's existing approach to drugs for AIDS. It is scientifically and medically sensible and consistent with legal precedents. And the FDA's intransigence in not conforming to its own policies argues the necessity for a change in the *law*, which the agency will find it more difficult to ignore.

The Clinton administration's opposition to HR 3199 has been intense and less than principled. Assistant Secretary of Health and Human Services Phil Lee has dismissed the bill and anything resembling it as nothing more than veto-bait. He has concentrated his efforts, instead, on weakening the already emasculated Kassebaum (Senate) bill, as a possible basis for compromise. David Kessler registered FDA's opposition to the House bill in a nine-page statement, "The Impact of the House FDA Reform Proposals,"[18] that is a revealing example of the lengths to which Kessler will go to protect the status quo at FDA. It is a masterpiece of untruths, misrepresentations and fear-mongering.

Kessler asserted that "FDA would be forced to approve new drugs using summaries of safety data prepared by drug companies." Untrue. The bill merely clarifies that rather than reviewing the voluminous raw data from clinical trials, often running to hundreds of thousands of pages, condensed, tabulated or summarized data often will be adequate. Agency reviewers would have access to additional materials if they were requested by FDA supervisory officials.

Kessler cited the example of a drug called Dilevalol, which he said was approved in Japan, Portugal and England on the basis of data summaries, while Americans were spared because "the FDA medical reviewer noted in the raw data evidence that some patients had severe liver injury." Another untruth. The record shows it was the *company*, Schering-Plough, that identified the toxicity and ultimately withdrew the application for U.S. approval.[19]

Kessler claimed that the legislation would weaken the effectiveness standard for drugs and that FDA would be forced to approve a new use for a drug on the bases of "anecdotal evidence of effectiveness" and "common use by physicians (with no objective evidence)." The result, he concluded, would be "the unnecessary pain and suffering patients would undergo until they were given an effective treatment." More untruths.

The reality is that FDA's policies have made life progressively more dangerous for patients and difficult for physicians. A 1995 survey of clinical oncologists found that nearly two-thirds believe FDA has "hurt their ability to give the best possible care to a patient on at least one occasion," and over 1 in 10 believe this has happened "frequently." Three-quarters "oppose FDA restrictions on off-label [as yet unapproved uses of approved drugs] information," and 60% believe these make their job more difficult.[20]

The congressional reforms would make it easier for health professionals to learn about legitimate new uses of drugs. HR 3199 would ease FDA's censorship of scientific and medical information, by permitting the legitimate dissemination of information via textbooks and articles from peer-reviewed journals. Equally important, because it goes to the basic issue of how new uses are sanctioned for an already-approved drug, the legislation would permit retrospective evidence from clinical research to constitute the basis for approving additional uses. This is a sensible and much-needed alternative to expensive, time-consuming and often duplicative prospective studies.

Flawed regulatory policies—screened from public awareness by public relations campaigns and subterfuges—are not uncommon at federal agencies. These policies afford little protection to human health or the environment and, overall, may exert a negative effect. They have huge costs and divert resources from other, legitimate public and private sector endeavors. They often discourage research and development and the marketing of new, innovative products. They breed well-deserved cynicism about government's motives. They benefit primarily one special interest: the bureaucrat's.

Clinton administration officials, including David Kessler, have treated regulatory reform as a political game—one in which sick Americans are treated as pawns. They ignore that the ultimate goal of government is to benefit consumers and that innovative new products save lives and ease suffering. With their "reforms" prescribed by spin doctors, only legislative change is likely to provide the needed structural, procedural and cultural improvements that are needed (chapter 5).

THE FDA'S *VOLTE-FACE* ON BIOTECH FOOD POLICY

Several lines of evidence argue both the insincerity of Clinton administration officials about FDA reform and the absence of FDA's commitment to improvement. First, the administration has sought repeatedly to institute new regulation of biotechnology foods that would reverse the paradigm of FDA's successful 15 year policy on biotechnology regulation.

On orders from Gregory Simon of the vice president's staff (see chapter 2), in 1993 FDA actively planned to require registration of all new biotechnology foods. This new policy directly contradicted the agency's widely-praised 1992 policy statement which specified that new biotechnology foods would be treated in the same manner as other, similar foods. The rationale for the policy reverse stemmed from a gratuitous "controversy" over biotech foods that was fueled by criticisms from a small and familiar cadre of antibiotechnology activists (see Chapters 1 and 5).

In addition to case by case, every—case review and mandatory labeling of all foods derived from rDNA-modified plants, the activists demanded *clinical trials* of these products. Their rhetoric was characteristically rich in cliches and misrepresentations that have become the stock-in-trade of biotechnology's critics. FDA had considered and rejected the same demands during the discussions about the 1992 food policy statement; FDA's conclusion was that there was no possible public health gain from the significant new costs that the extreme measures would impose on the government, industry and consumers.

Increasing the costs is, of course, a hidden part of the activist agenda. (Higher costs are particularly effective at impairing the competiveness of a technology, because they discourage consumers, producers and researchers, alike (see below)). Another strategy is to use FDA's regulatory stridency as a signal to the public that there is something fundamentally different and worrisome about biotech foods. The activists try to create a self-fulfilling prophecy: they argue that we need regulation because consumers are apprehensive; when consumers become apprehensive because the products are stringently regulated, the activists say we still need more regulation to assuage their concerns....

FDA and the Clinton White House officials have generally been receptive to extremists' demands. They have seemed more concerned about placating activists than ensuring access by consumers to the fullest array of improved foods and providing American companies opportunities for product innovation and business development.

Scientific experts persistently criticized the prospective food-registration requirement (which was not announced officially but was frequently mentioned in speeches by FDA officials).[21] In 1995, the new Republican Congress made pointed inquiries about the policy. When industry withdrew its support, Kessler was eventually compelled to abandon the idea, at least for the time being. The proposal was gone but not forgotten.

In July of this year, FDA began surreptitiously to circulate the word about new requirements. without any announcement or fanfare, a seven page document, "Foods Derived from New Plant Varieties: Consultation Procedures" went out to state officials.[22] In it, FDA adopts a pretense that the new policy is applicable to all new plant varieties not just those produced with the new biotechnology. But the agency's intent is transparent: oversight of the consultation process rests with the *Biotechnology* Evaluation Team (BET), and the degree of detail requested from the plant's developer would only be available for those crafted with new biotechnology. FDA is betting that Congress is too busy with the November election to notice that the agency is again up to its old tricks.

FDA's recent attempts to overregulate biotech foods reverse its own 15 year old guiding principle for the agecy's oversight of biotechnology: regulation should focus on real risks and should not turn on the use of one technique or another. That tenet has provided effective oversight for more than 1,000 new biotechnology products, including drugs, vaccines, diagnostic tests and foods. As noted above, as recently as May 1992 the FDA formally reiterated this policy for foods, affirming that new biotechnology foods would be treated no differently from those produced with other techniques.

In several ways the policy will discourage the application of biotechnology to foods. The data requirements are substantial: FDA lists nine categories of obligatory information and the detail is far greater than would be required for food products made with less precise, less sophisticated techniques. The policy will entail significant costs for the government and industry and, by extension, the public: according to the FDA's description of the new regulatory scheme the Biotechnology Evaluation Team will always consist of no fewer than six FDA staff, drawn from different parts of the agency. There will be endless and conflicting demands for information about each product, causing delay and uncertainty among manufacturers. The new regulatory regime will put every biotechnology product squarely in the sights of anti-biotechnology actvists: the results of consultations with industry will be available on the Internet.

The bottom line is that the policy will discourage research on more varied, appetizing and nutritious foods—research that has given us low saturated fat oils, seedless grapes, tangelos and the like. American farmers and food processors will be less competitive and consumers will be deprived of new choices. Moreover, the new approach challenges the very foundation of the agency's longstanding policy of regulating new biotechnology products on the basis of product risk or intended use, rather than genetic process.

Labeling for Biotech Foods?

Thwarted in their desire to get FDA to require clinical trials of biotechnology-derived foods, activists have retreated to demanding labeling that would inform consumers when the techniques of new biotechnology were used in a food's manufacture. The ostensible rationale for such a requirement is that information is power and that consumers can never know too much about the products they buy. Especially for foods, the more information the better.

But that's not necessarily true. A message can mislead and confuse consumers if it is irrelevant, unintelligible or crafted to tell only part of the truth. Moreover, a requirement for labeling carries added production expenses and raises costs to both producers and consumers that can constitute a barrier to the development of and access to new products.

To serve the consumer best, regulation should focus on genuine risks and require only that information about a food's origin or use that is relevant to safety and that supports genuinely informed choice. Mandatory labeling of all biotech foods would achieve none of this.

FDA's long-standing common-sense approach to food labeling has been that label information must be both accurate and "material." FDA does not require a "product of biotechnology" or "genetically engineered" label for foods from plants or animals that have been improved with rDNA techniques. In the 1992 food policy statement, FDA clarified that labeling is required "if a food derived from a new plant variety differs from its traditional counterpart such that the common or usual name no longer applies to the new food, or if a safety or usage issue exists to which consumers must be alerted."[24]

The 1992 policy statement emphasized that, as for other foods derived from new plant varieties, no premarket review or approval is required unless the characteristics of a new biotechnology product raises explicit safety issues. The policy emphasized that these safety issues could be raised by food from new plant varieties however they were created. The safety issues include the introduction of a substance new to the food supply (and, hence, lacking a history of safe use), increased levels of a natural toxicant, changes in the levels of a major dietary nutrient, and transfer of an allergen to a milieu where a consumer would not expect to find it (say, peanut protein transferred to a potato).

The FDA clarified that if a new food raises any of these safety issues, it could be subject to FDA regulations for premarket testing, product labeling, or removal from the marketplace. The FDA cited the example of new allergens in a food as a possible material fact whose omission could make a label misleading. The agency reiterated that the genetic method used in the development of a new plant variety is *not* considered to be material information because there is no evidence that new biotech foods are different from other foods in ways related to safety. Therefore, the FDA said that product labeling will not be required to include the method of development of a new plant variety. Biotech foods would not be required to be labeled as such.

The 1992 FDA policy statement has already been tested. A scientific article in the March 1996 issue of the *New England Journal of Medicine* reported that allergenicity common to Brazil nut proteins was transferred into soybeans by genetic engineering and was readily identified by routine procedures.[24] In effect, this report validates the FDA policy. The plant breeder, Pioneer Hi-Bred International, was required to and did consult with the FDA during product development. During the course of consultation and subsequent analysis, the allergenicity was identified. Confronted with the dual prospects of potential product liability and the costs of labeling all products derived from the new plant variety, the company abandoned all plans for using the new soybeans in consumer products. Not a single consumer was exposed to or injured by the newly-allergenic soybeans. In what might be considered a "positive control," the system worked.

The approach the FDA took in its 1992 policy statement is consistent with scientific consensus that the risks associated with new biotechnology-derived products are fundamentally the same as for other products. Dozens of new plant varieties modified with traditional genetic techniques (such as hybridization and mutagenesis) enter the marketplace every year without premarketing regulatory review or special labeling. As discussed in chapter 1, many of these products are from "wide crosses" in which genes have been moved across natural breeding barriers (and without rDNA techniques). None of these plants exists in nature. None requires or gets a premarket review by a government agency. (Safety tests by plant breeders consist primarily of taste and appearance and, in the case of plants with high levels of known intrinsic toxicants—

such as tomato and potato—measurement of the levels of certain alkaloids.) Nonetheless, plants from wide crosses have become an integral, familiar and safe part of our diet: wheat, corn, rice, oats, black currants, pumpkins, tomatoes and potatoes.

There are other reasons why special regulations and labeling requirements are often not in the best interest of consumers. As food producers know well, requiring a label can add significantly to the production costs of certain foods, particularly those that are produced from pooled fresh fruits and vegetables. To maintain the accuracy of labels, rDNA-modified fruits and vegetables would have to be segregated through all phases of production—planting, harvesting, processing and distribution—which adds costs and eliminates economies of scale. Added production costs, in turn, raise consumer prices and disadvantage products in the highly competitive, low profit-margin marketplace of processed foods.

Superfluous labeling requirements for new biotech products would constitute, in effect, an unwarranted tax on the use of a new, superior technology. The requirement would exact excess costs and reduce profits to plant breeders, farmers, food processors, grocers and others in the distribution pathway. The power of regulatory disincentives is such that this burden could virtually eliminate new biotechnology tools from food research, development and production. How far would food label requirements extend? Would special labels be required for foods such as pizza or burritos containing cheese made with new-biotech-produced chymosin (rennin), for chickens raised on feed from new-biotech-manipulated corn, and for beef from cattle vaccinated with a new biotech vaccine?

An analysis of the economic impacts of a labeling requirement for new-biotech-foods by the California Department of Consumer Affairs (CDCA) predicted that the additional costs would be "substantial," and that "while the American food processing industry is large, it is doubtful that it would be either willing or able to absorb most of the additional costs associated with labeling biotech foods." [26] The analysis concluded that "there is cause for concern that consumers will be unwilling to pay even the increased price for biotech foods necessary to cover biotechnology research and development, much less the additional price increases necessary to cover the costs associated with labeling biotech food."

The CDCA assessment implies another outcome of unwarranted but compulsory labeling. Overregulated and, therefore, overpriced biotech products would be limited to upscale, higher-income markets. Wealthier consumers would be able to pay more for the improved products, while the less affluent would simply do without them.

The FDA's 1992 food biotech policy is scientifically defensible and favors the public interest, but the FDA has shown by its recent policy changes on biotech foods and feeds a willingness to sacrifice scientific principles, bow to political pressure and accommodate activists' whims.

ESSAY 2
REGULATION AT THE EPA: NEITHER THE BEST NOR THE BRIGHTEST

The arcana of government policy often seem to have an uncertain relationship to everyday life. Occasionally, public events illustrate the real-world impacts of flawed policies. In the world of biotechnology regulation there is a surfeit of examples. They include the killing frosts in South America that precipitously raised world coffee prices several years ago and the bioremediation used on the 1989 *Exxon Valdez* oil spill in Alaska. Each represents opportunities for applying the new tools of biotechnology to serious problems. Each also represents opportunities lost to regulatory interference.

THE ICE-MINUS FIASCO

In the 1980s scientists at the University of California and in industry took an interesting approach to reducing frost damage to crops. They knew that a harmless bacterium called *Pseudomonas syringae* lives on many plants and contains an "ice nucleation" protein that actually promotes frost damage to plants. This occurs in a mechanical way, the protein serving as the initiator of the growth of ice crystals that damage the fruit and leaves of plants, including fruits and vegetables. In the presence of the bacterium, ice forms at higher temperatures. The scientists reasoned that spraying a variant, or mutant, of the bacterium that lacks the ice-nucleation protein (and which is, therefore, said to be "ice-minus") on plants might take up all the available space and displace the common "ice-plus" variant, thereby reducing frost damage.

In fact, they did test some spontaneous ice-minus variants but were plagued by the problem of the bacteria reverting to resume production of the protein. The gene splicing techniques of the new biotechnology seemed an ideal tool: scientists could irreversibly remove the entire gene for the ice-nucleation protein, rendering the bacterium permanently ice-minus. Despite some experimental success with this innovative approach, government regulations were to pose insurmountable barriers.

The EPA astonished the scientific and regulatory communities by classifying as a *pesticide* the obviously innocuous ice-minus variant of the bacterium, which was to be sprayed on potatoes and strawberries in order to test its ability to prevent frost damage. The EPA reasoned that the natural ice-plus bacterium is a "pest" because its ice-nucleation protein serves as a nidus for ice crystal formation and that a mutant intended to displace it is therefore a pesticide—a convoluted rationale that could lead one to regulate outdoor trash cans as a pesticide because litter is an environmental pest. This is an unfortunate example of regulators' torturing the intention of a statute in a way that is inappropriate and was never intended, in order to achieve their own ends.

The EPA said that an intensive and comprehensive review of the ice-minus bacterium was necessary because it was the first recombinant DNA-manipulated organism to be "released to the environment." But this was a specious rationale. It ignored the wide consensus that the use of rDNA techniques, per se, does not confer risk. Also, it disregarded the fact that this ice-minus bacterium differed from its wildtype cohorts by having a gene *deleted*: because genes are constantly deleted and mutated in nature, this genetic construction represents virtually no difference from what is already found commonly in nature. Finally, as to the presence of recombinant organisms "in the environment," a study funded by the EPA had already confirmed the substantial incidental release from laboratory "containment"—approximately 10^8 organisms/day/technician—of recombinant bacteria that occurs from standard research labs.[1] Thus, a vast and varied "release experiment" involving thousands of laboratories and millions of discrete new genotypes of recombinant microorganisms had been in progress for a decade, with no untoward results.

There was virtual unanimity among scientists, including those within the EPA, about the safety of the test of the ice-minus organism, but just because it was recombinant DNA-manipulated the field trial was subjected to an extraordinary, lengthy and burdensome review. (By contrast, research with spontaneous ice-minus variants had required *no governmental review of any kind.*) And even when approval for field testing had finally been granted, the agency continued its heavy-handedness by conducting elaborate and unnecessary monitoring of the actual field trials in the guise of performing research on risk.[2] (Making the situation more ridiculous, as they performed this superfluous monitoring of obviously harmless experiments—at significant expense to the taxpayers, of course—EPA officials were attired in super-high-containment moon suits more appropriate for protection against biological warfare.)

Two of the major stated purposes of the EPA's monitoring "experiment" were to: (1) "determine distributions of the GEMs [Genetically Engineered Microorganisms] on and off the plot during the meteorological conditions encountered during the spray and postspray periods" and (2) "provide recommendations for evaluating fate and transport of recombinant organisms for future field releases."[3]

EPA considered irrelevant the already available field trial data on wildtype and spontaneous mutants of *P. syringae*. The EPA's vast expenditure on monitoring obtained no data of scientific interest, and the agency's second stated goal could not be accomplished by such an experiment. Even in the ideal case, the data would only be meaningful for *P. syringae* in the particular environments where the tests were conducted. For example, the data would have no application to soil bacteria or other microorganisms transmitted by animal or insect vectors.

It is noteworthy that Dr. Steven Lindow, the University of California, Berkeley microbiologist who proposed the experiment, had previously performed field trials with spontaneous mutants of *P. syringae* that were phenotypically identical to the recombinant ice-minus organism—and these trials had required no government review, no notification and no special safety precautions of any kind. (Moreover, data from these field trials were actually submitted as part of the proposal to test the recombinant ice-minus strain.) Regulators were similarly unconcerned about field trials with the wildtype *P. syringae* tested for its ability to enhance the production of artificial snow at ski resorts. Without any scientific rationale for it, the use of recombinant DNA techniques was—and continues to be—the EPA's preferred regulatory trigger.

Despite the demonstration that the bacteria were safe and effective at preventing frost damage, the combination of onerous government regulation and the huge expense of doing the experiments discouraged additional research. The product was never commercialized. The supply and price of citrus, berries, coffee and other crops remain a hostage to the vagaries of killing frosts.

The ice-minus fiasco is not an isolated example. At around the same time, the Monsanto Company proposed a small-scale field trial that was scientifically interesting and potentially important—control of a corn-eating insect by a harmless soil bacterium, *Pseudomonas fluorescens*, into which scientists had cloned a protein from another, equally innocuous bacterium. Despite the unanimous conclusion of the EPA's external scientific advisory panel and other federal agencies (I wrote the FDA's opinion) that there was virtually no likelihood of significant risk in the field trial, the EPA refused to permit it.

Two aspects of this situation are noteworthy: the field trial would not have been subject to any government regulation at all, had the researchers used an organism with identical characteristics but crafted with less precise "conventional" genetic techniques; and Monsanto's response to the rejection was to dismantle its entire research program on microbial biocontrol agents. This program would have developed biological agents to replace chemical pesticides—an express goal of EPA Administrator Carol Browner.

EPA has had a chilling effect on the entire sector of biocontrol R&D.

Bioremediation: Another EPA Casualty

Bioremediation, the detoxification of wastes with living organisms, is another casualty of regulatory disincentives. A 1994 report in the journal *Nature* described modest success in cleaning up the beaches fouled by the 1989 *Exxon Valdez* oil spill in Prince William Sound, Alaska.[4] However, the bioremediation techniques that were employed—pouring fertilizer on the beach to stimulate the growth of any bacteria that were there—represent science and technology worthy of the 19th century. Consider the observation of William Reilly, the EPA administrator at the time of the accident, "[w]hen I saw the full scale of the disaster in Prince William Sound in Alaska...my first thought was: Where are the exotic new technologies, the products of genetic engineering, that can help us clean this up?"[5] Good question.

Innovative products of the new biotechnology for bioremediation remain on the drawing boards because researchers and companies have been intimidated by regulatory barriers and disincentives. EPA officials have tried for a decade to issue final regulations for field trials with microorganisms—including those for bioremediation. They always return to the wrong answer: proposals that discriminate against microorganisms created with high-precision recombinant DNA technology while exempting research organisms crafted with any other technique (*vide infra*).[6]

The Clinton administration has encouraged and promoted these scientifically indefensible approaches, enabling some regulations actually to issue in final form; and the quality of regulatory policy has deteriorated during its tenure. Recombinant DNA technology applied to bioremediation or the development of microbial pesticides is almost nonexistent. Indeed, because regulatory submissions for bioremediation research with recombinant DNA-manipulated microorganisms are overly burdensome and are thrust by regulatory procedures into the public domain, and because of the uncertainty of ultimate marketing approval, the U.S. bioremediation industry has largely restricted itself to work with naturally occurring organisms. But because of the nature and complexity of the substances involved in most spills and toxic wastes, naturally occurring organisms are often suboptimal or inadequate for the job. The new biotechnology provides additional tools for the microbiologist.

Agricultural biotechnology and bioremediation, among other industrial sectors, represent scientific and commercial opportunities sacrificed to the EPA's regulatory disincentives. This is the legacy of bad public policy—that is, government "business as usual" in the Clinton administration. That is not to say that all of it arose on their watch; it did not. However, under Vice President Gore's influence, in particular, both the product and process of public policy formulation have deteriorated. No expense, no cost in U.S. industries' lost innovation or competitiveness, no sacrifice of consumer products has been as important to regulators as theoretical (and often, imaginary) improvements in the health and safety net and the furtherance of a "green" agenda. It was only the voters' installation of the Republican-controlled 104th Congress—exercising both its appropriations and oversight roles—that applied the brakes to the agencies. Skeptics should ask themselves, "How often did I hear the term 'regulatory reform' from the administration or its agencies before the 1994 mid-term elections?"

Regulators as a Special Interest Group

The battle over a group of amendments to the EPA's FY96 appropriation bill provides some interesting object lessons. House Republicans fought a little-heralded but titanic battle during the summer of 1995 to amend the EPA's appropriation bill to trim regulatory excesses. Congressman James T. Walsh (R-NY) introduced an amendment that would seem unlikely to ruffle feathers: "That none of the funds appropriated under this heading may be used to exclusively regulate whole agricultural plants subject to regulation by another federal agency." It would also have inserted the following language at the appropriate place in the committee's report:

> The Committee notes that genetically engineered plants are subject to significant regulatory scrutiny, including by the Food and Drug Administration to ensure food safety and the Department of Agriculture to avoid release of plant pests or other environmental hazards. The EPA has proposed to broaden its regulation under the Federal Insecticide, Fungicide and Rodenticide Act (FIFRA) to whole plants that have increased pest resistance developed through biotechnology. The Committee directs EPA to curtail its regulation of genetically engineered plants to avoid redundant regulation and minimize burdens on beneficial research and development. Specifically, such regulation should be limited to application to agricultural plants which contain a pesticidal substance that does not naturally occur in nature or has been regulated under FIFRA when applied externally to plants.[7]

The amendment was adopted by the full House Appropriations Committee on July 18th, 1995, but, in debate in the full House it encountered a firestorm of opposition from certain segments of industry (see chapter 5). On July 31, a 210-210 procedural vote left the amendments intact. On the following day, President Clinton decried the congressional action, blaming lobbying by nefarious "special interests." He promised to defy them by vetoing the bill.

What Mr. Clinton does not know, or will not acknowledge, is that regulatory agencies have themselves become special interests. (Some EPA officials were even lobbying members of Congress on the Walsh amendment, an apparent violation of federal law.) The president's lapse is no different from that of many advocates of big government, who often seem perplexed by federal agencies' poor performance.

During the first two years of the Clinton administration, the EPA announced several final or preliminary policies for regulating various products that are genetically engineered and that have pesticidal properties.[8] These policies represent the culmination of a 10 year effort to come to regulatory terms with rDNA technology. Advances in understanding of the scientific bases of risk during that period allowed ample opportunity for scientific principles to drive EPA's policies.[9] Yet, even as recently as November 1994, the EPA has cast into federal regulatory policy an anachronistic approach that targets the rDNA *techniques* used to create these organisms. The EPA has steadfastly discounted others' conclusions and recommendations concerning the safety of rDNA methods and organisms. Instead, the EPA has maintained its course of singling out rDNA-modified organisms for special consideration which translates into burdensome and unnecessary regulatory reviews.[10]

Scientifically defensible, viable policy alternatives have been implemented by the National Institutes of Health (NIH) and Centers for Disease Control (CDC) for laboratory work with pathogens, and proposed by the National Research Council and others.[11] They are discussed in chapter 5.

Flawed Policies: Regulating Processes Instead of Risks

The EPA has a lengthy history of policy formulation based on considerations other than scientific predictions or measures of risk related to environmental protection. For example, in the late 1980s, in response to a widespread media campaign waged primarily by the Natural Resources Defense Council, the EPA pressured apple growers to abandon the use of the plant growth regulator Alar, an agricultural chemical that permits apples to ripen uniformly and increases yield. The EPA's capitulation to environmentalists' demands conflicted with the agency's own scientific findings.[12] But faced with a public relations barrage against Alar, EPA Assistant Administrator John Moore issued a statement asserting that "there is inescapable and direct correlation" between exposure to UDMH (the primary degradation of product of Alar) and "the development of life threatening tumors," and that therefore the EPA would soon propose banning Alar.[13] However, as the EPA admitted separately, there was no data to support a finding of carcinogenicity. Moore "urged" farmers who were using Alar to stop.[14] Coming from a senior federal regulator, that is akin to an armed mugger "urging" the victim to relinquish his wallet.

During the Alar episode, one of Moore's senior subordinates, a lawyer, attempted to intimidate the members of an advisory panel because their opinion differed from his own.

> Apparently, the EPA officials had expected the SAP [Scientific Advisory Panel] to rubber-stamp its decision [that Alar or UDMH was a carcinogen]. When it did not, Uniroyal [the manufacturer of Alar] officials were jubilant. But after the meeting, Steven Schatzow, then director of EPA's Office of Pesticide Programs, herded SAP members into his office. The angry Schatzow demanded, 'How can you do this to us?' After a heated exchange with the scientists, he concluded, 'Look, I can't tell you what to do, but you might like to think about this one again.' The scientists were stunned by such flagrant interference, and all refused to back down.[15]

Environmentalist demands appear likewise to have influenced the EPA's approach to regulating products of the new biotechnology. The EPA's final rule for "genetically engineered" microbial biocontrol agents under FIFRA (the pesticide statute), was published on September 1, 1994.[16] While this regulation represented an opportunity for scientific principles to drive regulatory policy, the EPA, with the active collaboration of other parts of the Clinton administration, instead adopted a highly centralized, intrusive approach that once again targets techniques rather than high-risk organisms, environmental results or outcomes. Under the new regulation the EPA would regulate innocuous organisms such as *P. syringae* or *Rhizobium* that contain, for example, the *E. coli lac* ZY genes as a metabolic marker or luciferase as a visual marker, just because rDNA techniques have been used.

Characteristically, the FIFRA rule does not identify the use of rDNA techniques forthrightly as its regulatory trigger. Rather, the rule uses circumlocutions. It targets "deliberate genetic modification," then defines "deliberate" as "directed," and finally equates "directed" with the use of molecular techniques. In the end, the rule regulates phenotypically identical organisms differently because different genetic techniques are used. This is a typical EPA rule: bad policy derived from bad science and running to a hundred thousand words of impenetrable bureaucratese composed, revised and reviewed during tens of thousands of civil-servant-hours.

For a decade the EPA has offered proposals on the scope of biotechnology regulation—that is, which products or experiments are subject to regulatory requirements. They always target the newest and most precise molecular genetic manipulation techniques despite scientific consensus (reviewed in chapter 1) that process or technique per se does not correlate with environmental risk. The use of the new gene-splicing techniques should not constitute a trigger for regulation. Recall the National Research Council report's observation that, compared to the imprecision of classical techniques of gene transfer or modification, "with organisms modified by molecular methods, we are in a better, if not perfect, position to predict the phenotypic expression [of organisms in field trials]."[17]

EPA's biotechnology policies have conflicted with official federal policy that was developed with EPA's formal agreement. That policy stipulates that regulation of biotechnology products should be "risk-based," "scientifically sound," and focused on "the characteristics of the biotechnology product and the environment into which it is being introduced, not the process by which the product is created."[19] The EPA statements of policy often use the appropriate buzzwords while subverting the unmistakable intent of the scientific principles.

The EPA has ignored bona fide risk considerations in favor of unsubstantiated fears expressed by special interest groups. One of the EPA's convoluted arguments holds that "newness," in the narrow, literal sense, is highly correlated to risk, and that because rDNA techniques can easily be used to create new gene combinations, rDNA manipulations therefore "have the greatest potential to pose risks to people or the environment."[20] The result of this specious reasoning is extensive, expensive, case-by-case screening of virtually all small-scale field trials of rDNA-manipulated microorganisms, while tests of similar—or even phenotypically identical—

organisms manipulated by other techniques are exempted. (EPA offers scant improvement through its exemption in the new rules of recombinant organisms whose genome has only undergone deletion or rearrangement—obvious examples of self-cloning).

The EPA's policy exempts field trials of "naturally occurring" organisms and organisms manipulated by chemical or radiation mutagenesis or by transduction, transformation or conjugation. That includes organisms likely to foul waterways or pose other serious environmental risks. The EPA's long-standing (pre-rDNA) policy was to exempt small-scale field trials of all microorganisms and chemicals. Important precedents include the EPA's policies under FIFRA and TSCA, neither of which has traditionally required case-by-case reviews. Under FIFRA, all small-scale (defined as less than 10 acres) field trials of microorganisms were considered automatically not to require the EPA's review and were exempt from regulation. There is a similar but more vague exemption under TSCA of chemicals or microorganisms for "small quantities solely for R&D."[21]

While these approaches can hardly be said to be risk-based, there was a certain logic. Small-scale experiments generally are not of great safety concern. Under these exemptions, R&D was performed with thousands of variants of microorganisms, for purposes as varied as pest control, frost prevention, enhanced artificial snowmaking, promoting the growth of plants, mining, oil recovery, bioremediation and sewage treatment.

Except for microorganisms manipulated with rDNA techniques, that policy remains intact.

Bad Policies, Bad Decisions

The EPA's process- or technique-based approach under FIFRA is only one example of scientifically flawed regulation and consequent negative impacts on R&D focused on improving agriculture and the environment. The EPA has begun a process that would require the review of a whole category of products that haven't heretofore been perceived as requiring any regulation—whole plants genetically modified (with rDNA techniques) for enhanced pest resistance.[22]

As discussed in chapter 1, plant varieties have long been selected by nature and bred by humans for improved resistance or tolerance to external factors that inhibit their inherent survival and productivity. These factors include insects, disease organisms, herbicides and environmental stresses. All plants contain resistance traits, or they would not survive. Thus, the issue is not the presence or absence of pesticidal properties, but degree. Moreover, there is no evidence to suggest that the degree of pest resistance is correlated with risk to the environment.

Plant breeders, farmers and consumers possess extensive experience with crops and foods that have been genetically modified for pest resistance.[23] In recent decades, so-called "alien" genes have been transferred widely across natural breeding boundaries by chromosome substitution or by embryo rescue techniques to yield commonly available food plants.[24] Most often, the plant breeders seek resistance to plant pests, such as nematode or bacterial canker resistance in tomato, and late blight or leaf roll resistance in potato.[25] These plants are, in fact, "genetically engineered," although not rDNA-manipulated. Their produce is commonly available at the local supermarket or farm stand.

EPA is moving FIFRA inexorably toward case-by-case regulation of rDNA-manipulated pest- or disease-resistant plants. Ignoring genuine risk considerations, regulated field trials would include those with wheat or corn with an additional single gene for chitinase or one of the other newly discovered disease-resistance genes. It is ironic and disturbing that EPA is heaping discriminatory regulatory burdens on the new molecular techniques just as they are yielding a "bumper crop of disease resistance genes" that may be "the biggest thing since the discovery of chlorophyll."[26]

EPA's proposal was vetted by an extramural scientific advisory panel whose report warned that any such regulatory policy "will have a profound effect on future plant breeding programs" and that an unwise choice of regulatory scope will constitute a potent disincentive to research and commercialization. The panel was unable, however, to make the connection between this observation and its ultimate recommendation to limit regulation to rDNA-manipulated plants. Ignoring both scientific considerations and the impending disincentives to research and development (which the quotation above suggests was of concern to them), the panel members based their recommendation largely on the dubious, 20-year-old precedent of the NIH rDNA Guidelines (which was by then obsolete and no longer applicable to the oversight of plants) and on "a public perception that there are risks."[27]

Historically, plants have not been subjected to risk assessment and management in the manner applied to pesticides. Oversight for the safety of food, including food from new plant varieties developed with the aid of recombinant DNA (rDNA) techniques (genetic engineering), is provided by the Food and Drug Administration (FDA). Similarly, oversight for protection of the environment from the release of plants genetically modified by rDNA techniques is provided by the U.S. Department of Agriculture (USDA).

To complement and extend this regulatory framework, the Environmental Protection Agency (EPA) proposed, on November 23, 1994, to regulate the inherited traits of plants that confer resistance to pests under statutes developed for chemical pesticides: Proposed Policy: Plant-Pesticides Subject to the Federal Insecticide, Fungicide and Rodenticide Act (FIFRA) and the Federal Food, Drug and Cosmetic Act (FFDCA). Based on this proposal, the substances produced by plants for their defense against pests and diseases, together with the necessary genes for production of these substances, would fall under a new category of pesticide termed plant-pesticide, subject to the statutes of FIFRA and FFDCA.

Concern among plant, food and microbiological scientists for the scientific basis and unnecessary regulatory burdens of this proposed policy was so great that a consortium of scientific societies convened a group of scientists to examine the scientific basis of the proposed rule and develop principles for appropriate oversight of the inherited traits in plants for resistance to pests and diseases. Plant breeding has been used with great success for most of this century to develop plant varieties with inherited traits for resistance to pests, including viruses, bacteria, fungi, nematodes, insects and mites, with an overwhelming record of safety.

Eleven scientific societies representing more than 80,000 members endorsed the consortium report. Each of these societies shares a common mission to disseminate scientific information. The scientists representing these societies bring diverse knowledge and experience about plants, foods from plants, plant pests and diseases, plant defense mechanisms, and techniques to develop and use new disease- and pest-resistant varieties of plants for agriculture, forestry, landscapes, gardens and other environments.

Scientific concerns

The primary scientific concern with the proposed rule is the creation of a new category of pesticide, called "plant-pesticide," solely for the purposes of regulation under existing statutes. EPA proposes to designate as "plant-pesticides" all substances responsible for pest resistance in plants, as well as the genes needed for production of these substances. Under its proposed policy, however, EPA singles out for possible registration as "plant-pesticides" only those traits introduced into plants using rDNA techniques. This happens as follows: under the statutes of FFDCA, the EPA establishes tolerances for the formulations and intended uses of pesticides. Under the new category of "plant-pesticide," EPA proposes to exempt from requirement for a tolerance those pest-defense substances and the genes necessary for the production of these substances, if they evolved naturally or were transferred to the plant by traditional plant breeding methods. Thus, tolerances for "plant-pesticides" would be established specifically for pest-defense traits that could not be transferred to the variety by traditional breeding methods and were therefore transferred to the variety using rDNA methods. A tolerance would also be required for varieties of plants developed by traditional breeding methods if one of the parents carried an inherited trait for pest resistance originally transferred to that parent by rDNA methods.

The consortium reached the following conclusions:

- It is scientifically indefensible to regulate the inherited traits of plants for pest and disease resistance under statutes developed specifically for chemical pesticides applied externally to plants;
- All plants are able to prevent, destroy, repel or mitigate pests. Further, all plants are resistant to most potential pests (susceptibility is the exception), although the actual mechanisms of pest defense are complex and the roles of any specific substances remain largely unknown;
- While pest resistance can be determined by specific genes, the ability to respond to and resist pests is a characteristic of the plant and cannot be separated for regulatory purposes from the plant itself and
- Evaluation of the safety of substances in plants should be based on the toxicological and exposure characteristics of the substance and not whether the substance confers protection against a plant pest.

The consortium supports the conclusions of independent studies by the National Academy of Sciences published in 1987 and 1989, which states that any risk associated with genetic modification of an organism is the result of the characteristics of the organism with its new trait(s) and not the method used to transfer genes for the new trait to the organism.

Economic and public policy concerns

The consortium recognizes the role of the federal government in assuring environmental and public safety of plants new to the environment and of plant products used as food. However, the EPA proposal designating the inherited traits of plants for resistance to pests as "plant pesticides" would, if implemented, have several negative consequences for agriculture and consumers. The EPA-proposed rule will:

- Erode public confidence in the safety of the food supply by sending the message that all plants contain "pesticides;"
- Discourage the development of new pest-resistant crops, thereby prolonging the use of synthetic chemical pesticides;
- Increase the regulatory burden for those developing pest-resistant varieties of crops, while also increasing federal and state bureaucracy;
- Limit the use of rDNA technology for the development of pest-resistant plants to those developers that can pay the increased costs associated with additional regulation and deny the benefit of this technology to applications for niche markets likely to be developed by small companies and public plant breeding programs;
- Handicap the United States in competition for international markets because of U.S. government policy that new pest-resistant varieties, or products from these varieties, be identified as containing their own "pesticides" and
- Limit the use of valuable genetic resources and new technologies to improve crop protection from pests and diseases.

Recommended principles for oversight

Several principles are appropriate to guide federal oversight for novel kinds of plant varieties intended for use in agriculture, gardens, urban landscapes or other managed ecosystems:

- Federal oversight of plants should be based on accepted standards of practice for recombinant DNA research and field-testing of plants. These include nongovernmental peer review and recommendations/guidelines provided by the U.S. National Academy of Sciences, the Organization for Economic Cooperation and Development and the U.S. Department of Agriculture;

- Regulatory oversight should focus on high-probability risk rather than hypothetical or unrecognizable risk and should be sufficiently flexible to keep pace with new scientific developments;
- The level of risk of a plant variety to the environment or human safety is determined by the characteristics of the plant, not by novelty or initial lack of familiarity, the source of the gene(s) that produce a pest-defense substance or initiate a pest-defense reaction, nor the method by which a gene for pest defense is transferred into the variety;
- Genes and the substances encoded by them that confer resistance characteristics to plants are not the equivalent of "pesticides" as defined by FIFRA and
- GRAS, the FFDCA concept of "generally regarded as safe," applies only to foods and food additives, but is an appropriate concept for environmental risk management. Mechanisms should be developed for conferring the environmental equivalent of GRAS status to new varieties of plants as we gain experience and familiarity.

Recommendations

The major recommendations of the report are:

- Ensure that federal oversight for plants with inherited traits for resistance to pests is based on the Principles for Oversight stated in this document;
- Strengthen and make greater use of nonregulatory oversight mechanisms developed for plants, such as plant variety review boards and scientific peer review that have served effectively during this century and
- Simplify the federal oversight of plants genetically modified to express novel traits, including those for pest defense, to be consistent with the risk-based policies of the FDA for food safety and the USDA for environmental protection.

The community of plant, food and microbiological scientists is dedicated to advancing safe and sustainable methods for enhancing the quality and productivity of plants in managed and natural ecosystems and welcomes the opportunity to work with the EPA and other federal agencies responsible for oversight.

Assertions of widespread public concerns are not supported by reliable data.[28] Even if they were, public misapprehensions about risk would be a flimsy basis for costly regulatory policies. Of greater concern, perhaps, is the fact that perceptions of public perceptions were presented as a rationale for unscientific regulatory policies recommended by a *scientific* advisory panel.

The only other rationale offered by the panel was based on the erroneous assertions that only with rDNA techniques is it "possible to make novel genetic combinations never before possible," and that novelty is synonymous with risk (*vide supra*; see also chapter 5). Extending this logic, one could argue that the EPA should institute case-by-case review of the presently unregulated new genotypes produced by plant breeders using conventional genetic techniques. Each year an individual breeder of corn, soybean, wheat or potato tests in the field as many as 50,000 new genotypes.[29] Many of these new variants are the outcomes of interspecies or intergeneric wide crosses that transcend natural breeding barriers, and their genomes are, at best, poorly characterized. Not surprisingly, no one has suggested that these field trials require government review.

The rationale for EPA's plant-pesticide proposal lies so far outside scientific norms that it stimulated unprecedented action by the scientific community. Eleven major scientific societies convened a meeting in January 1996, specifically to draft a report opposing the policy. Because the report, approved by the elected leadership of each of the professional scientific societies, illustrates so comprehensively and eloquently many of the points I have made about EPA's biotechnology regulation, the report's Executive Summary is quoted here in its entirety [see box, pp. 112-116].

A Scientifically Challenged Agency

The EPA's pursuit of policies that regulate the technique rather than the characteristics of the product is only one manifestation of the agency's documented neglect of science advice applied to policy formulation, a deficiency that has been criticized by eminent extramural scientific groups. An expert panel commissioned by then-EPA Administrator William Reilly reported in 1992 that: (a) "The science advice function—that is, the process of ensuring that policy decisions are informed by a clear understanding of the relevant science—is not well defined or coherently organized within EPA;" (b) "In many cases, appropriate science advice and information are not considered early or often enough in the decision making process;" (c) While "EPA should be a source of unbiased scientific information...EPA has not always ensured that contrasting, reputable scientific views are well-explored and well-documented;" and most damning of all, that (d) "EPA science is perceived by many people, both inside and outside the Agency, to be adjusted to fit policy. Such 'adjustments' could be made consciously or unconsciously by the scientist or the decision-maker."[30]

The panel was charitable. I found the EPA to be by far the most scientifically-challenged agency that I encountered in almost two decades of public service, during which I interacted frequently with many government departments and agencies. The EPA's ability to propose and apply flawed scientific assumptions or paradigms to regulatory policy is intimately related to the manner in which the agency handles its advisory process. Not infrequently on policy issues related to the new biotechnology, the EPA maneuvered scientific advisory panels on predetermined courses; when scientists on those panels offered independent perspectives, they angered federal officials. For example, when University of California microbiologist Dennis Focht, an academic member of the EPA's Biotechnology Science Advisory Committee, observed in a letter to the committee's chairman that a policy decision to regulate on the basis of genetic technique rather than according to risk was based on nonscientific considerations,[31] he was subjected to a written rebuke from EPA Assistant Administrator Linda Fisher. A lawyer, she chided this distinguished scientist on his inability to "provide the Agency with [an] unbiased assessment of the scientific issues at hand," and, in effect, invited him to resign from the committee.[32] This unseemly treatment of a scientific advisor is not an isolated incident; another EPA senior administrator's outrageous behavior towards extramural scientific advisors, which occurred during the agency's deliberations on Alar, is described above. Incidents like these subvert the ability of scientists to contribute fully to public policy decisions.

The Focht example is a situation in which an extramural advisor, an eminent academic scientist, was genuinely committed to providing rigorous, objective and apolitical advice to federal regulators. This is the usual case at agencies like the National Institutes of Health and FDA; and even though advisory committee members may be (and often are) NIH grantees (and NIH and FDA are sister agencies, located within the Department of Health and Human Services), that is not considered to constitute a conflict of interest. Advisors usually provide narrow scientific expertise by reviewing and ranking grant applications, expressing opinions about research areas ripe for additional funding, or evaluating the results of clinical trials. The federal agency does not have a large vested interest in any particular decision by its extramural committees.

A different and perverse situation frequently prevails at the EPA. Instead of narrow scientific questions, the biotechnology-related committees are often asked for opinions on *policy* issues. More to the point, they are often asked, in effect, to rubber-stamp a course of action that directly benefits and is preferred by the EPA.

As discussed above, the EPA consistently has chosen policy directions that serve bureaucratic ends (such as larger budgets and regulatory empires), while disadvantaging academic and most industrial research. This places extramural advisors in a position that is, at the least, uncomfortable, and at worst, frankly conflicted. It is noteworthy that at the time that the EPA was proposing and its advisory committees were recommending scientifically indefensible and regressive policies, some chairmen and members of EPA's Biotechnology Science Advisory Committee were receiving substantial agency funding.

In addition to manipulating advisory committees, I found that EPA staff had a greater capacity for chicanery in day-to-day interactions with Congress, other agencies, governments and international organizations than I observed anywhere else. This included intriguing with congressional staff, leaking confidential government documents to "green" nongovernmental organizations (NGOs), divulging U.S. negotiating strategies to foreign governments, and falsifying the "reporting cables" required by the State Department to summarize international conferences (*vide supra* and chapter 5).

In an effort to elevate EPA's scientific profile, in 1989 the agency brought on board Dr. William Raub (former NIH deputy director) as the senior science advisor to the Administrator. Raub was known to be a smart, savvy and collegial scientific administrator. Nonetheless, the EPA staff proceeded to make his life miserable. From the beginning, they ignored him when they could. When they couldn't, they sent him drafts of important documents too late for a meaningful review—often just days before a court-ordered deadline for an agency action. EPA Administrator Browner excluded Raub from her inner circle and finally replaced him with a less-threatening lower level EPA staffer.[31]

The EPA appears to rank near the bottom in quality, among the United States' vast governmental bureaucracies. Some of the programmatic and policy deficiencies can be ascribed to career civil servants with their own agendas, who manipulate EPA-inexperienced political appointees (most often lawyers). A pertinent example is Dr. Elizabeth Milewski, an EPA mid-level manager who has long had primary responsibility for biotech policies in the Office of Pesticides and Toxic Substances. At a 1991 interagency meeting that I attended, she announced huffily that the EPA could not accept a certain scientifically based policy because "our constituency won't stand for it."

As a government official, I understood "our constituency" to be the American taxpayers and consumers who gave us their trust and treasure, but Milewski was referring to a much narrower and more vocal constituency: antibiotechnology activists at the Environmental Defense Fund, National Wildlife Federation, Union of Concerned Scientists and Greenpeace—who think of government regulation as something with which to bludgeon technologies, industries or companies they dislike.

A particularly vivid example of the confluence of bad science, flawed policy and execrable regulators' performance occurred at the EPA in September 1995. A group of anonymous agency scientists publicly opposed and successfully derailed the agency's approval of a genetically engineered soil microorganism, for stimulating plant growth.

The microorganism in question, *Rhizobium meliloti*, belongs to a class of common soil-dwelling bacteria that form nodules along the roots of legumes—alfalfa, peas and soybeans, for example—and "fix," or capture, nitrogen from the air. In essence, the bacteria convert atmospheric nitrogen to a form that serves as a plant nutrient—in a way that is cheaper and more environment-friendly than, for example, applying nitrogen fertilizer. Such bacteria have been cultured and used in agriculture for more than a century, and cultures of *Rhizobium* have been sold in the United States since 1895. They are considered perfectly safe for humans, animals and plants that come into contact with them. This *Rhizobium* variant: (a) should not have been subject to EPA regulation at all; (b) was over-evaluated and over-analyzed by EPA reviewers; (c) was reviewed incompetently by the EPA's advisory committee and (d) still has not been approved for marketing (as of October 1996). In scientific terms, this is like a positive control experiment that doesn't work—that is, even when conditions are intentionally "gamed" to give a positive result, you don't get it. Lamentably, that's true of much of what the EPA does.

Via one of its typical scientific and semantic circumlocutions, EPA managed to capture *Rhizobium* for regulation under TSCA because it contains DNA from sources of organisms that are classified in different genera. This is rather like saying that recovering a metal screw from a nuclear submarine and using it to repair lawn furniture should trigger a special and extraordinary regulatory regime, because of the different sources of materials—or, at the very least, the furniture should be subject to the level of regulatory scrutiny appropriate to the submarine. Characteristically, EPA's policy makes no sense whatsoever.

It is difficult to fathom how this microorganism could be construed to fall under TSCA at all, but at most, the *Rhizobium*—engineered for enhanced nitrogen fixation, in order to boost alfalfa yields—should have been subject to no more than the EPA's customary treatment of microorganisms under the TSCA regulations; as discussed above, these regulations exempt tests of chemicals or microorganisms when a field trial involves "small quantities solely for R&D." As to the marketing of organisms, TSCA traditionally regards all (non-rDNA-manipulated-) organisms as "natural" and, therefore, not subject to case by case, premarket review.

For microorganisms containing mixed sources of DNA (a backdoor way to capture recombinant DNA-manipulated organisms), however, EPA eliminated the exemption, capturing most rDNA-manipulated microorganisms but promising "expeditious reviews." In my experience, that's about as likely as an August snowstorm at the EPA's Washington D.C. headquarters—as the *Rhizobium* story illustrates.

Consider this excerpt from a news item about EPA's review of *Rhizobium,* in *Bio/Technology*:

> In mid-1994, officials of the U.S. Environmental Protection Agency (EPA, Washington D.C.) ran into some last-minute opposition just when they seemed poised to approve the first application for the commercial-scale release of a genetically-engineered microorganism. Last month [January 1995], members of EPA's Biotechnology Science Advisory Committee (BSAC) analyzed and set aside several objections to that application...for a genetically improved strain of the nitrogen-fixing bacterium, *Rhizobium meliloti,* known as RMB PC-2. However, the members of BSAC did not resolve several of the "subtler" issues surrounding RMB PC-2, which involve potential environmental impacts of the product. This leaves Research Seeds [the producer] in commercial limbo for at least the next few months and possibly out of luck for another growing season.[33]

With the product apparently nearing approval in September of 1995, a document began circulating under the auspices of a group called Public Employees for Environmental Responsibility (PEER) that purported to represent the views of "several EPA scientists" that "overeager to promote biotechnology, EPA has either deliberately ignored or actively suppressed concerns raised by staff and independent scientists."[33] EPA pro-biotech? The absurdity of that suggestion alone speaks to the reliability of the group.

The document goes on to describe a few implausible scenarios, such as toxicity of the recombinant *Rhizobium* to humans (roughly as likely as a squirrel monkey becoming King Kong by eating a bottle of hormone tablets); and what they call "The Frankenstein Effect"—the microorganism converting clover into an aggressive, kudzu-like plant. (For EPA scientists, maybe not a surprising speculation—but hardly in the realm of possibility; see chapter 1.)

The adoption by the EPA of risk-based approaches to oversight would have been a win-win proposition—and not at all difficult to achieve (see chapter 5). The advantages would have been decreased direct government spending on regulation, stimulation of public and private sector R&D by removing the burden of regulatory disincentives, and reassurance to the public about the essential equivalence of new biotechnology and other more traditional techniques.

However, the risk-based approach presented clear administrative and budgetary disadvantages for EPA and this ultimately prevented the agency from adopting it. Instead, EPA followed the course of bureaucratic self-interest, and its approach to biotechnology regulation has been a make-work project extending over more than a decade. The agency has drafted and published numerous regulatory proposals, collated and responded to public comments, conducted lengthy negotiations with the Office of Management and Budget and other agencies, promulgated and implemented regulations and, in the process, created and populated new boxes on the EPA organizational chart.

All of this may be good for the care and feeding of EPA officials, but it has been bad for agricultural research and for the development of much needed biological pest management strategies (see also the discussion of USDA policies, below).

Regulatory disincentives, increasingly enshrined in final regulations, will continue to deter researchers and companies from biological control strategies that could substitute safer genetically engineered microorganisms or plants for chemical pesticides. Innovations that may not provide sufficient financial return to offset the inflated costs of testing and registration are especially vulnerable.[35] By limiting available technological choices, the EPA's regulatory philosophy and policies are likely to damage, rather than protect, both agricultural research and the environment.

Superfund: A Nonbiotech Boondoggle

The Congress-Clinton administration budget impasses and partial government shutdowns of the winter of 1995 raised questions in the popular media and public mind about the distinction between "essential" and "nonessential" government services, as the latter shut down for a time. The shutdowns caused various analysts (of whom I was one) to reflect on the seminal question that makes big-government advocates cringe: which government "services" would we be better off without? Which major programs should be axed completely?

A logical place to start is the federal regulatory agencies, which are expensive, intrusive "special interests." As I have discussed elsewhere in this chapter, many of the EPA's biotechnology policies are excellent examples—but not isolated ones.

Superfund, an EPA program intended to clean up ("remediate," in EPAspeak) and reduce the risk of toxic-waste sites, is another good example. This program was originally conceived as a short-term project—$1.6 billion over 5 years to clean up some 400 sites (by law, at least one per state and, not coincidentally, about one per congressional district). But it has grown into one of the nation's largest public works projects: $30 billion spent on almost 1300 sites. Superfund, more formally the Hazardous Substances Trust Fund, is worse than merely an unwise or inefficient policy. It is a prime example of a large government program that inflicts violence on Americans.

Various studies have attempted to evaluate the effect of Superfund's massive and costly cleanups, but the results are uncertain. Putting that another way, no beneficial results have been demonstrable after the expenditure of 30 thousand millions of dollars. On the other hand, Superfund projects have caused a great deal of harm.

San Jose State University economics professor J. Paul Leigh has analyzed the occupational hazards of environmental cleanup projects. He concluded that the risk of fatality to the average cleanup worker—a dump-truck driver involved in a collision or a laborer run over by a bulldozer, for example—is considerably larger than the cancer risks to individual residents that might result from exposures to unremediated sites.[36] (And consider that cancer risks are theoretical estimates over many years or decades, while worksite fatalities occur during the much shorter time of the cleanup.)

Leigh's studies imply that there are three important factors the EPA should take into consideration in directing a cleanup. First, worker fatality risks tend to increase as the desired levels of cleanup increase, since more soil excavation and transportation are required to make the site cleaner. In other words, if the EPA requires the removal of 99% of the waste instead of 90%, vastly more work and more time at risk are necessary.

Second, the baseline risks at contaminated sites are often small because of the small number of people who live near sites. Third—and this follows from the first two points—the regulators must compare, or balance, the risks to different groups. In the official records of decisions at many Superfund sites, however, the possibility of dangers to cleanup workers is not even mentioned.

Supreme Court Justice Stephen Breyer, in his *Breaking the Vicious Circle*, argues further the need to compare risks:

> the regulation of small risks can produce inconsistent results, for it can cause more harm to health than it prevents. Sometimes risk estimates leave out important countervailing lethal effects, such as the effect of floating asbestos fibers on passersby or on asbestos-removal workers... Sometimes the regulator does not, or cannot easily, take account of offsetting consumer behavior, as, for example, when a farmer, deprived of his small-cancer-risk artificial pesticide, grows a new, hardier crop variety that contains more 'natural pesticides' which may be equally or more carcinogenic.[37]

Even former EPA Administrator William Reilly has suggested that Superfund's risk-assessment paradigms are flawed. In a speech at Stanford University in January 1994 while a visiting lecturer, he noted that basing cleanup on exaggerated worst-case scenarios could lead to excessive costs:

> Superfund has relied on different exposure assumptions from other EPA programs, though it conducts its risk assessments similarly. The risks it addresses are worst-case, hypothetical present and future risks to the maximum exposed individual, i.e., one who each day consumes 2 liters of water contaminated by hazardous waste. The program at one time aimed to achieve a risk range in its cleanups adequate to protect the child who regularly ate liters of dirt... And it formerly assumed that all sites, once cleaned up, would be used for residential development, even though many lie within industrial zones. Some of these assumptions have driven cleanup costs to stratospheric levels and, together with liabilities associated with Superfund sites, have resulted in inner-city sites suitable for redevelopment remaining derelict and unproductive. The consequence, in New Jersey and other areas, has been to impose a drag on urban redevelopment in the inner city, and to push new industry to locate in pristine, outlying sites.[38]

Clinton administration officials have equated budget cuts at agencies like EPA with putting an arbitrary "price" on life. What they fail to recognize is that unnecessary regulations also put a price on life by draining limited government resources away from serious risks, in order to control negligible or even imaginary risks. To deny that the cost is relevant to regulatory policy is to deny economic reality. As the Cato Institute's Jerry Taylor observes,

> Money spent on Superfund risks is money not spent on something else, including the ability to protect public health in other ways, reduce poverty, improve public safety, or even the intangible (but very real) benefits one gets from disposing one's income as one likes. According to the EPA's (extremely dubious) estimates, for example, $1 million spent on Superfund saves approximately 2.5 years of life. But $1 million spent on breast cancer screening saves 300-700 years of life. Similarly, $1 million spent on cervical cancer screening saves 700-1500 years of life.[39]

An EPA scientist, Carl Mazza, explained during a 1995 conference in Washington D.C. sponsored by Harvard University's Center for Risk Analysis that the agency is aware that Superfund policies often conflict with risk analysis but "political considerations" don't permit rational decisionmaking. I'm sure that taxpayers and cleanup workers will find that comforting.

Leaving aside the fine points of risk analysis, most of the costs of Superfund actually end up going to lawyers. Usually, I would find that outrageous, but it's certainly preferable to spend money on Mercedes-Benzes for lawyers than for killing bioremediation workers.

There are other indirect and subtle perils of government overregulation. Money spent on implementing and complying with regulation (justified or not) exerts an "income effect" which reflects the correlation between wealth and health, an issue popularized by the late political scientist Aaron Wildavsky. It is no coincidence, he argued, that richer societies have lower mortality rates than poorer ones, and to deprive members of society of wealth is to enhance their risks.[40]

Wildavsky's argument is correct: Wealthier individuals are able to purchase better health care, more nutritious diets and generally less stressful lives. Conversely, the deprivation of income itself has adverse health effects, including an increased incidence of stress-related problems, including ulcers, hypertension, heart attacks, depression and suicides.

It is difficult to quantify the relationship between the deprivation of income and mortality, but academic studies suggest, as a conservative estimate, that every $7.25 million of regulatory costs will induce one additional fatality through this "income effect."[41] The $4 billion annual cost to society of Superfund—most of it transferred to consumers as higher prices on chemical and petroleum products—would, therefore, be expected to cause more than 500 deaths per year. These are the real costs of "erring on the side of safety," the mantra which Vice President Gore and EPA chief Carol Browner invoke to justify regulatory overkill.

Programs like Superfund that afford little or no protection to human health or the environment are not uncommon at agencies like EPA. They have huge costs and divert resources from other legitimate public and private sector endeavors. They breed well-deserved cynicism about government's motives. Like FDA's policies on drug advertising and promotion, they benefit primarily one special interest: the regulators themselves.

My purpose in this chapter is not to review comprehensively the life and times of the EPA. However, on the basis of my own observations, experience and research on three major EPA programs—FIFRA (pesticides), TSCA (toxic substances) and Superfund—there is good reason to question whether the quarter-century experiment with a free-standing Environmental Protection Agency has been a success. Given the murkiness of its statutes, the content of its regulations, the chronic neglect of scientific principles as the basis for decision making, the misuse of its advisory committees and the quality of protoplasm of its staff, I am not optimistic about fixing it (but see chapter 5).

ESSAY 3
BIOTECHNOLOGY REGULATION AT THE USDA: THE SEAMIER SIDE OF GOVERNMENT SCIENCE POLICY

This treatment of biotechnology regulation at the USDA focuses primarily on the *why*, rather than the *what* of USDA's biotechnology regulation. Readers who want more about mainstream biotechnology oversight at the USDA—that is, regulation of veterinary biologicals and field trials of transgenic plants by the Animal and Plant Health Inspection Service (APHIS)—can find elsewhere the regulatory minutiae,[1] the APHIS party line,[2] and the real story.[3]

Self-Interest and Shifting Loyalties

While I was a mid-level civil servant responsible for formulating biotechnology policy at the FDA in the 1980s, various governmental panels discussed *ad nauseam* the appropriate "scope" of biotechnology regulation. The government's choice of scope—whether to focus on product or process—would have important implications. Historically, federal agencies largely focused on products. The advent of genetic engineering brought furious cries for regulation of the genetic technique, or process, for the first time. Inadequate regulation of new processes could compromise human health or environmental protection. Conversely, new and excessive regulation would create disincentives to using the new techniques.

Prestigious scientists and scientific groups provided critical guidance, concluding that new biotechnology, which uses rDNA techniques that mimic and speed up nature's own movement of genes between organisms, should be freed of much regulatory red tape. In 1987, Dr. Charles Hess, then Dean of the School of Agriculture at the University of California, Davis, wrote that

> there are valid scientific reasons to believe that products developed through recombinant DNA technology will pose problems no greater than biological agricultural agents developed by more traditional processes, such as breeding and cell culture...in contrast, genetic changes produced by natural mutation, breeding and cell culture are often not well understood.[4]

He went on to criticize governmental regulatory approaches that focused on the new techniques, observing that "the government for the first time is imposing significant regulatory restrictions on products with no known hazards."[5] His words echoed the consensus of independent national and international groups everywhere, and presaged the conclusions of the National Research Council's (NRC) landmark comprehensive report on the subject (see chapter 1).[6]

However, the message fell on many deaf ears. As the two preceding essays in this chapter have already indicated, despite the clear consensus among scientific experts, some government officials sought to build bureaucracies around rDNA-manipulated organisms. This, then, was the crux of the controversy over the scope of regulation.

The FDA had apprehended from the time of its initial contact with the new technology in the late 1970s that the new techniques were a refinement, or improvement, over earlier techniques for genetic manipulation. Therefore, the FDA decided to treat biotech products no differently from similar products made with conventional technology. Officials at the USDA argued an opposite position. They maintained that public opposition to the new biotechnology mandated that the scope of regulation should encompass all products made with the new techniques, regardless of actual risk. Not coincidentally, this would require a large new regulatory bureaucracy to be established at USDA.

Discussions and disagreements over "scope" dragged on. In 1989, President Bush's senior scientific advisor for the biological sciences, Dr. James Wyngaarden, suggested that I submit a scholarly paper describing the scientific perspective on the issue to a prestigious journal, where it would be peer-reviewed and critiqued. I and several eminent coauthors from within and outside the government did submit such a paper to the journal *Science*. That's when the fireworks began.

As a courtesy, one of my coauthors sent a copy of the submitted paper to Dr. Hess who had by then been appointed USDA Assistant Secretary for Science and Education. We were surprised to find that Hess's views had dramatically and suddenly changed since arriving in Washington. Hess was livid at the proposal in our paper even though he had only recently supported precisely the same views. When he moved from the academic community to USDA, his allegiances and his view of the stakes changed. If our ideas were accepted, the USDA's plans for an expanded regulatory empire were doomed.

With bureaucratic turf at stake, Dr. Hess actively sought to stop the publication of the paper. He had the chief of staff of the Secretary of Agriculture contact me through his counterpart at the FDA and demand that I withdraw the paper. I refused. Dr. Hess then tried to dissuade the *Science* editors from publishing the paper. However, the editor-in-chief took umbrage at this unseemly pressure and said that had there been any doubt about whether to accept and publish the paper, Dr. Hess' coercion alone would have induced him to do so.[7]

The paper was published in *Science*[8] and in 1992 the federal government adopted endorsed its basic tenets as official policy.[9]

This unsavory tale illustrates how easily an academic can shed the ethos of the scientific community, when his self-interest is at stake. There are similar examples at many government agencies to illustrate this theme. At the USDA two other sagas are relevant. One pertains to Terry Medley, head of the Animal and Plant Health Inspection Service's (APHIS) Biotechnology, Biologics and Environmental Protection program; and the other, to Alvin Young, director of the department's Office of Agricultural Biotechnology. Hess was a political appointee, Medley and Young are civil servants. In all of them, we see the corrupting influence of self-interest on the formulation of public policy, and the way it distorts officials' behavior.

Humpty Dumpty and Biotechnology Policy

The saga of Terry Medley highlights some of the distortions created by the existing system of bureaucratic incentives, disincentives, rewards and punishments. Over the past decade Medley has assumed a great deal of power. He has ignored or resisted official U.S. biotech regulatory policy, conducted what amounts to his own foreign policy, and attempted to impose his idiosyncratic view of biotechnology regulation on foreign governments.

Medley consistently eschewed scientific principle in the debate over whether the federal regulation of new biotechnology should target either new techniques or the risk-related characteristics of products. While high level policymakers were working during the 1980s to implement a product-based regulatory approach in various U.S. regulatory agencies–in contrast to Medley's process-based one–Medley was on the move. He traveled to Bonn, Paris, Brussels and other capitals, preaching regulatory "harmonization" and assuring foreign governments that he would deliver a USDA policy similar to the process-based approach being implemented by the European Union (then called the European Community), even if it conflicted with White House policy.

In 1987, under Medley's direction APHIS promulgated Plant Pest Act regulations specifically for organisms manipulated with rDNA techniques and containing any amount of DNA from a plant pest. Organisms were regulated even in the face of compelling evidence of the extremely low probability of transforming a benign organism into a pest.[10] Such an outcome would be about as likely as making the family cat a man-eater by giving it a single gene from a lion. In field trials and even in transport (say, from one university to another), organisms were subjected to a federal permitting process that included a comprehensive and expensive environmental assessment. In fact, all of the thousands of permits for field trials or transport issued under these regulations have been for organisms of negligible risk.

By mid-1992, it was clear that the USDA's policies were significantly discouraging agricultural research and development, so the Bush administration sought to rationalize the regulations. Medley responded with a scheme to replace permits with notification under circumstances so narrowly defined, however, that only a few companies or university researchers would have benefited. He provided the illusion of regulatory reform but left the plight of researchers largely unchanged and his own regulatory domain undiminished.

The Bush administration modified the proposal to make it more risk-based and scientifically defensible. This new plan made a notification process applicable to greater numbers of research trials and to a wider spectrum of plant species. It would have been a greater stimulus to innovative research. It would also have made redundant much of Medley's regulatory empire. Not surprisingly, he argued strenuously against it.

Medley's strategy was not an uncommon one inside the Beltway (the circular highway that encircles metropolitan Washington D.C.). A lawyer, he argued to other lawyers that the scientific considerations supported his position. To scientists, he argued that scientific evidence was of secondary significance in policymaking: of greater importance were public perceptions of risk (no matter how misguided), reassuring consumers, and ensuring that U.S. policies conformed to those of our trading partners (no matter how scientifically flawed or economically regressive). The alternative to his approach, Medley warned, would be militant consumers and European and Japanese trade barriers to U.S. products. The administration's top policy makers were unimpressed. The proposal, as improved, was published on November 6, 1992.[11]

By the time the comments on the regulatory proposal were received and analyzed, a new presidential administration had assumed office. Clinton policy makers removed the improvements that the Bush administration had introduced into Medley's scheme. In order to protect his bureaucracy, Medley had bypassed his own supervisors and the Office of Management and Budget, and intrigued with his friend Greg Simon in the office of the new vice president, Al Gore. On March 30, 1993, the final rule was published—one small step for R&D, a giant leap for bureaucratic empire and antitechnology activism. And hardly an auspicious beginning for Gore's "reinventing government."

Regulatory disincentives are potent, and Medley's impact on agricultural biotechnology research is no exception. Consider, for example, the effect of APHIS' regulatory policy on research activity.

The economics of the relationship between regulation and R&D has exerted an influence on the distribution of field trials among various kinds of institutions. Data from USDA show that only about 13% of field trials with transgenic plants have been conducted by publicly funded research institutions and that this fraction is decreasing (Figs. 3.1 and 3.2), despite the new Medley notification scheme. It is noteworthy, too, that the publicly funded institutions work on a much larger spectrum of organisms (Fig. 3.3).[12] Surveys have found that significant numbers of academic researchers in the United States are discouraged by USDA's regulations,[13] the U.S. National Association of State Universities and Land Grant Colleges (NASULGC) concluded that biotechnology research in public institutions has been severely impaired by the regulatory burdens of governmental policies.[14]

Medley, who bears much of the blame for APHIS' policies and their sequelae, has tried to reinvent himself as a born-again champion of science-based policy. In a July 14, 1994 memo to high-level officials at the FDA and the EPA, Medley accused the EU of controverting principles established in the General Agreement on Tariffs and Trade (GATT) which mandated that "health-related issues pertaining to food and agriculture should be decided on the basis of science rather than political or economical [sic] considerations." Medley, an advocate of science-based regulatory policy? Hardly.

It happens in government that as circumstances change, so do the bureaucrats' machinations. The EU policy has, in fact, the same

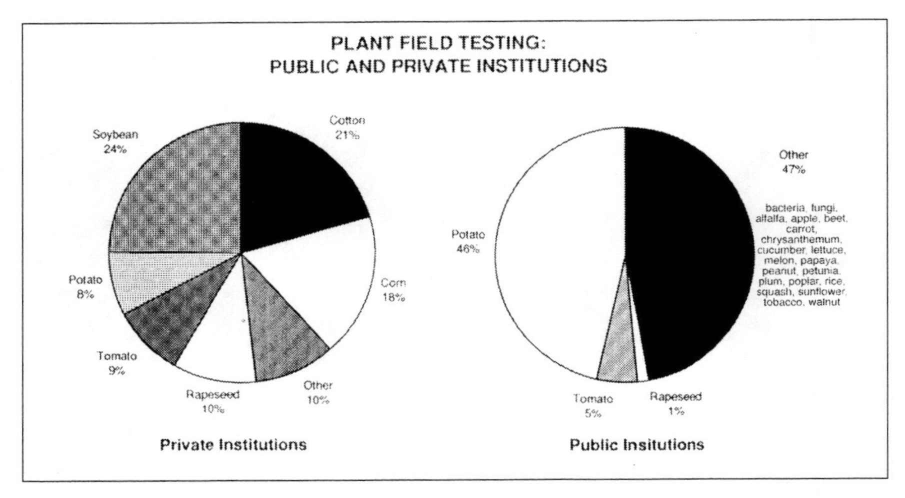

Fig. 3.1. Relative percentage of field trial permits and notifications approved by USDA Animal and Plant Health Inspection Service to applicants in commercial (privately-funded) and noncommercial (publicly-funded) institutions. Data provided by USDA. Reprinted by permission from Huttner SL et al, U.S. Agricultural Biotechnology: Status and Prospects. Technological Forecasting and Social Change, 50:25-39. Copyright 1995 by Elsevier Science Inc.

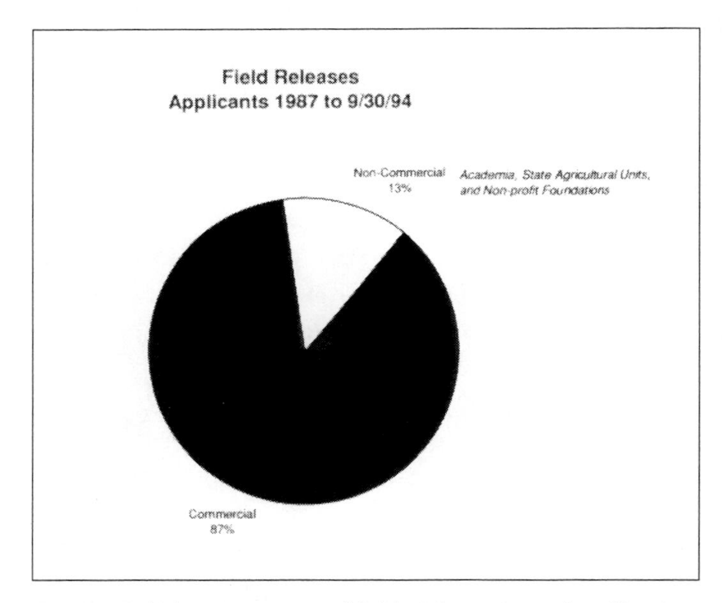

Fig. 3.2. Relative percentage of field trial permits and notifications approved by USDA Animal and Plant Health Inspection Service for specific crops to commercial (privately-funded) and noncommercial (publicly-funded) institutions during the period of 1987 to May 1994. Data provided by USDA. Reprinted by permission from Huttner SL et al, U.S. Agricultural Biotechnology: Status and Prospects. Technological Forecasting and Social Change, 50:25-39. Copyright 1995 by Elsevier Science Inc.

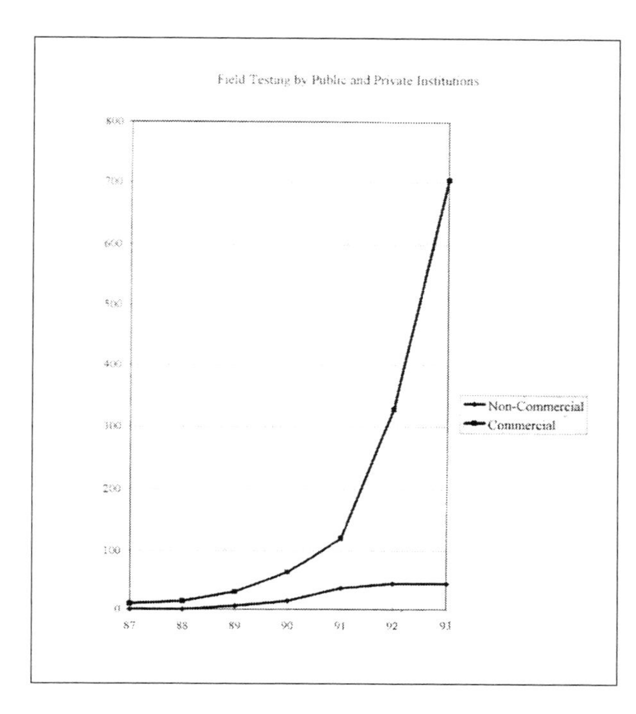

Fig. 3.3. The number of field trial permits and notifications approved by USDA Animal and Plant Health Inspection Service to commercial (privately funded) (upper curve) and noncommercial (publicly funded) (lower curve) institutions during the period of 1987-1993. Data provided by USDA. Reprinted by permission from Huttner SL et al, U.S. Agricultural Biotechnology: Status and Prospects. Technological Forecasting and Social Change, 50:25-39. Copyright 1995 by Elsevier Science Inc.

focus and emanates from the same scientifically flawed rationales that Medley consistently advocates in Washington (see chapter 4). Where the European proposal goes ever so slightly further than his own, Medley denounces it as unscientific. Like Humpty Dumpty in Lewis Carroll's *Through the Looking Glass*, he makes science mean just what he chooses it to mean—neither more nor less.

Unfortunately, while Medley and other regulators play their self-serving games, U.S. investors divert their money to R&D in other industrial sectors. Fewer agricultural biotech innovations enter or emerge from the R&D pipeline. Farmers and consumers are given fewer new choices. America's international advantage recedes. A level playing field is of little advantage when it is waist-deep in mud.

Medley is the consummate bureaucrat who knows how the system works, unfortunately for agricultural biotechnology. Perhaps it is unfair to blame Medley for taking advantage of a system whose incentives and disincentives, rewards and punishments, are "gamed" to reward such behavior.

In June 1996, Medley was appointed APHIS Administrator, a prototypic Clinton administration pick: he's political, cunning, a lawyer lacking technical expertise, and never met a regulation he didn't like.

USDA's "AIRhead" Program

A figure in contrast to Medley—or perhaps it would be more accurate to say, a more extreme manifestation—is Alvin Young, outgoing head of USDA's recently dismantled Office of Agricultural Biotechnology (OAB). The OAB was located on the organization chart in the part of USDA "opposite" Medley's regulatory domain—that is, in the huge empire concerned with agricultural "Science and Education." This part of USDA promotes agriculture by means of research stations, support to land-grant universities, competitive grants programs, extension services and the like. But Young, with a substantial budget, top-heavy staffing and no real functions, struggled for almost a decade to create a job for himself and his minions.

For several reasons, he found regulation of previously-unregulated research a tempting target, one made viable by continuing activist antics. First, case-by-case regulation (see below) of a robust research area is a veritable bonanza, in terms of bigger budgets and more staff. Second, the academic research community he would burden with new regulations is docile, disorganized, largely unaware of the implications of new regulation, and slow to react. Third, once regulation has been approved, the regulator has a virtual entitlement program. The regulatory agency needs adequate resources in perpetuity to keep churning out approvals, even if the entire apparatus is unnecessary and ill-conceived.

The single overarching goal of the OAB and its Agricultural Biotechnology Research Advisory Committee (ABRAC) was the development of the "ABRAC guidelines" for recombinant DNA agricultural research. Rather than offering its own expert and independent advice to the OAB, the ABRAC acceded to the OAB's vision of inherent, unique biotechnology risks and need for extensive government oversight.

In the end, the ABRAC guidelines were scientifically indefensible and not risk-based, being focused specifically on the use of a single technique. This placed them squarely in conflict with the scientific consensus about biotechnology described above and in chapter 1. The ABRAC approach was inconsistent as well with official U.S. government policy, as expressed in the so-called "scope policy,"[15] which reflected the consensus of the scientific community.

The guidelines were never approved by the Bush administration. It was only this sound policy decision that kept the OAB from jurisdiction over all field trials of rDNA-manipulated animals—and from massive increases in manpower and budget. Young's response to this setback was to send out thousands of copies of the guidelines bound and presented in a way that implied that they were en route to official sanction.

OAB's activities exemplify government gone wrong. They reflect a tenacious effort to build what Milton Friedman has called a government of the people, but by and for the bureaucrats. OAB officials wasted the time of untold professionals in the research and public policy communities. The promulgation of guidelines, standards for research, notices of policy and individual regulatory decisions required stakeholders in the American agricultural research enterprise to expend time responding to them—at least in writing, sometimes personally at public meetings.

OAB regularly organized and supported elaborate international conferences, ostensibly to share information about GMOs (genetically manipulated organisms). As discussed in chapter 1 (essay 5), however, the value of such "horizontally-organized" information is limited. In practice, the conferences perpetuated the notion that new biotechnology products present unique risks that warrant special regulation. U.S. taxpayers paid for conclaves in South Carolina, California, Germany and Japan. (For the 1994 conference in Monterey, California, Young transported virtually his entire office staff from Washington at taxpayers' expense, although none made a presentation.)

The conferences' focal point—field trials of GMOs— is a good example of a bureaucratic attempt to transform a scientific artifact into a sphere of influence. "Field trials of GMOs" is not a discrete category of scientific import or interest. It is, however, reminiscent of a parody published via the Internet in The mini-Annals of Improbable Research ("mini-AIR") called AIRhead Project 2000, which consists of a list of "studies, projects and products that involve the number 2000"—for example, Hella 2000, an automobile fog-lamp; The 2000 Club, a group of people organized by professional magician James Randi; and SP 2000 Kitty Litter, an inexpensive Finnish brand. OAB's less imaginative but equally irrelevant list included organisms ranging from bacteria and fungi to fruits, vegetables, cereals, trees and ornamental flowers.[16]

Congress cut $300,000 of OAB's current $500,000 budget in the FY96 appropriation, according to Undersecretary of Agriculture Karl Stauber.[17] He noted that the Congress pushed the USDA to shed advisory committees and specifically targeted the ABRAC, but it is uncertain whether its machinations will disappear completely: Stauber has "asked USDA managers to name functions that might be shifted to other parts of USDA."

Outright elimination of OAB is exactly the kind of action that Americans should expect—from the Congress, if USDA doesn't take the hint. It would reduce federal spending, remove an unnecessary burden from the research community, stimulate innovation and make government more credible. It would also establish an important principle: federal employees are entitled to be AIRheads, but not on the taxpayers' nickel.

More Biotech Problems at USDA

The USDA biotechnology saga described in this chapter offers very little to be upbeat about, and there is more I haven't touched upon.

One example is the National Biotechnology Impact Assessment Program (NBIAP), an expensive boondoggle that includes an extensive inventory of biotech-specific regulations, guidelines and other information. NBIAP is a classic example of the government undertaking a project that the private sector would and should perform, if there were a demand for it.

Another example is a set-aside of 1% of all of the USDA's competitive grants for "biotechnology risk assessment." This has a corrupting influence in several ways.

First, the sequestration of funding for this area implies (correctly) that the quality of risk assessment research is too low for proposals to compete on the basis of merit. Second, a report prepared by the National Science Foundation for the White House as long ago as 1985 concluded that:

> Environmental applications of biotechnology are not a new endeavor. Microorganisms have frequently been

> modified in the past by methods other than recombinant DNA and have been successfully and safely introduced into the environment... While risk assessment remains an inexact process, it provides a systematic means of organizing and interpreting a variety of relevant knowledge about the behavior of microorganisms in the environment.[18]

Thus, there is not a demonstrated need for risk assessment specific to biotechnology—or even any evidence that there is such a thing. Risk assessment research should answer questions that are of scientific interest and be funded on the basis of merit.

Finally, the existence of this special "pot" of money induces researchers to cast their proposals—of almost any sort—as "risk assessment," in order to be eligible for the set-aside. This is analogous to what economists would term a "market distortion."

Chapter 5 discusses both generic and specific suggestions for redressing bureaucratic behavior, but the USDA experience argues strongly for what is at the same time the simplest and most effective solution: structural changes that trim government activity, and therefore afford less overall opportunity for excess, distortion, cupidity and stupidity.

ESSAY 4
BIOTECHNOLOGY REGULATION BY THE NIH: AN UNCOMFORTABLE ROLE

The role of the venerable U.S. National Institutes of Health (NIH) as a biotechnology regulatory agency has been both complex and anomalous. This federal agency funds the majority of biomedical research in the United States and, in intramural laboratories on its sprawling campus outside Washington D.C., performs some as well.

The oversight of recombinant DNA (rDNA) research has been the NIH's only major regulatory role. Predictably, the quality of the science has generally been high, but science has often been poorly—or at the least, far too slowly—translated into effective public policy.

As noted in chapter 1, oversight of the new biotechnology by the NIH has been relatively enlightened, in the sense that early on its *Guidelines for Research Involving Recombinant DNA Molecules* (known colloquially as the "NIH Guidelines") evolved relatively rapidly to become less stringent. Over two decades, the NIH Guidelines have undergone numerous iterations.

At first, the Guidelines oversaw virtually all research in the United States that involved rDNA, and completely prohibited certain applications, such as human gene therapy and field trials. Later, the prohibitions were lifted, and these classes of experiments were permitted only after case-by-case approvals by the NIH (following review by its Recombinant DNA Advisory Committee, known as the "RAC").

Eventually, virtually all laboratory research was completely exempted from the jurisdiction of the NIH Guidelines, the oversight of most field releases was delegated to bona fide regulatory agencies, and only human gene therapy regulation was left under NIH's jurisdiction. The NIH's current role and the legacy of its unscientific approach to regulation are problematic.

On the positive side, the existence of the NIH Guidelines—conceived and constructed with extensive guidance from a small group of scientists—may have prevented the imposition of even less satisfactory legislative solutions to what might have been perceived as a "vacuum" in the oversight of genetic engineering.[1] From the beginning, however, the NIH approach was to treat recombinant DNA techniques as unique among all means of performing genetic manipulation—and society and the research and industrial communities have paid a high price for that invalid assumption. By creating a system of "self-regulation," scientists may have been acting out a kind of post-1960s urge to be socially sensitive or attempting to purge their collective guilt for the technological marvel that was the Manhattan Project. In the end, they not only got the science wrong (or at least incomplete, by not seeking the insights of specialists in fields like epidemiology, evolutionary biology, immunology and endocrinology before leaping ahead with regulation), but showed they had little comprehension of public policy. By crafting a technique-based regulatory mechanism—one emulated throughout most of the world—the NIH invented the wheel which would be rolled adroitly by generations of self-interested regulators and activist demagogues.

The current version of the Guidelines leaves human gene therapy—when and only when it involves recombinant DNA technology—the last real vestige of regulatory responsibility by the NIH. At the cutting edge of medicine, gene therapy has intermittently received intense attention when, in 1993, for example, the technique was successfully used to lower the astronomically high cholesterol levels of a patient with a genetic defect (familial hypercholesterolemia) and again in 1995, when a broad patent was granted for the basic technology. It is difficult to predict how important SHGT applied to genetic defects, cancer and cardiovascular disease will be in the near future. A conservative guess is that within two decades it may approximate what bone marrow transplantation is today: a standard—albeit expensive—first-line treatment for a few diseases, and the treatment of last resort for others. (Twenty years ago when I was in medical school, bone marrow transplantation was highly experimental, performed at only a handful of medical centers, and almost always unsuccessful.) But even this degree of progress is likely only in the absence of undue government interference. Overregulation of gene therapy will slow profoundly its testing and ultimate adoption.

The Overregulation of Gene Therapy

As of May 1996, 180 clinical trials of gene therapy—the insertion of therapeutic genes into the cells of human subjects, in order to correct genetic or acquired disorders—had been approved in a dozen countries. The disorders range from genetic deficiency diseases such as cystic fibrosis to AIDS and many solid tumors.

At present, however, excessive governmental regulation in many countries is creating major impediments to gene therapy's testing, researchers wasting time and resources. Without any scientific or medical basis for it, gene therapy is the most highly regulated procedure in medicine, and perhaps the most regulated of any in society.

Consider the situation in the United States. Like all other clinical trials of new drugs, gene therapy experiments are subject to the jurisdiction of the FDA and the Institutional Review Boards (IRBs) within the hospitals where the research will be performed. The FDA reviews the trial for medical and scientific soundness, and the local review board for ethical and medical considerations. However, because gene therapy involves rDNA technology, the clinical trial also falls under the jurisdiction of the NIH Guidelines. It is noteworthy that the Guidelines have no jurisdiction over other (nongene-therapy-related) vaccines or drugs, no matter how novel or potentially hazardous and no matter how they are made. The scope of NIH's oversight is, thus, not genuinely risk-based; rather, it's based on the arbitrary "category" of rDNA-mediated, human gene therapy. The NIH Guidelines' current jurisdiction over gene therapy is, in fact, a throwback to this "category" being one of the original prohibitions in the first version of the Guidelines. The NIH's continued jurisdiction over gene therapy protocols has important implications.

The NIH Guidelines mandate three extra, separate but duplicative NIH reviews: approval by Institutional Biosafety Committees (different from the IRBs that oversee all clinical research) at all institutions participating in the trial; the NIH's Recombinant DNA Advisory Committee ("the RAC," on which I served as FDA's representative from 1980 until 1993, longer than any other person); and the NIH director. The first and foremost defect of this regulatory scheme is redundancy. The NIH's reviews are internally redundant and they duplicate all of the others as well.

Second, few RAC members possess experience in critical areas: the manufacture and quality-control testing of pharmaceutical products; and the nuances of clinical trials, including what are considered the usual or appropriate risk/benefit considerations for clinical trials.

Lawyers, political scientists, ethicists and "public members" (whose *raison d'être* is often a political connection) are heavily represented on the RAC, and this should give one pause for thought. Is this inclusive, holistic, politically correct, New Age approach appropriate for deciding scientific issues? Should nonexperts perform heart transplants or decide which flu vaccines are safe? More to the point, should a political scientist or a plant biologist perform a primary review of a highly technical clinical protocol to determine whether, and the conditions under which, dying children should have access to a new therapy? (This has actually happened at RAC meetings.)

Not surprisingly, the NIH has sometimes held the treatment proposals to standards much different from those the FDA applies to all other experimental drug treatments. (With all its warts, the FDA is better than having amateurs make complex regulatory decisions.) The result has been that researchers are confused and spend their time on paperwork instead of experiments.

Third, the RAC holds only several brief meetings a year, which forces investigators (and patients) to wait for several months to have a proposal considered. Moreover, because the committee is not continually available to researchers, there exists no satisfactory mechanism for a researcher to obtain permission to make real-time important adjustments to the experiment, if problems occur or if other developments dictate changes in course.

In response to remonstrations by researchers and patients' groups, in September 1994 the NIH and FDA announced a "streamlining" of the regulatory process. The primary responsibility for gene therapy was to devolve to the FDA: investigators would submit only a single application and be informed within 15 working days whether they would have to undergo a single FDA review or a dual FDA-NIH review.[2]

This process finally went into effect about the middle of 1995, and by May 1996 approximately 40 protocols had been submitted that were subject only to FDA review. While this change does represent progress, it is too little, too late.

In July 1996, NIH Director Harold Varmus formally proposed terminating NIH's anomalous role as a (redundant) government regulator of clinical research and ceding its regulatory responsibilities to the FDA.[3] However, the proposal perpetuates the myths that rDNA technology is somehow different from other techniques and that human gene therapy is a discrete entity—one, moreover, that deserves special attention from NIH above and beyond that afforded other cutting-edge areas of research. Specifically, it proposes to establish a new Office of Recombinant DNA Activities (ORDA) Advisory Committee (OAC) that will "ensure public accountability for recombinant DNA research and relevant data;" a plan to hold frequent Gene Therapy Policy Conferences (GTPC) for "public discussion of the scientific merit and the ethical issues relevant to gene therapy;" and "continuation of the publicly available, comprehensive NIH database of human gene transfer clinical trials, including adverse event reporting."[4]

These proposals are variously gratuitous, unwise, or ill-conceived. NIH's existing mechanisms for convening consensus-building conferences—the Office of Medical Applications of Research, for example—are adequate for the scientific, medical and other aspects of gene therapy, just as they serve for a panoply of other subjects.

The proposal to maintain an NIH database, or inventory, implies, incorrectly, that "human gene–transfer clinical trials" is a group whose elements are sufficiently similar or scientifically interesting to make it of value. With this proposal, one might argue that NIH has come full circle. The NIH began oversight of biotechnology in the 1970s by circumscribing an ill-conceived and unscientific sphere of influence—recombinant DNA manipulation—as a "group" of activities worthy of special attention. Twenty years later, ignoring the mistakes of the past, NIH seems doomed to repeat them.

"Human gene therapy" is not, in fact, a genuine "category" that is amenable to an inventory. It is too disparate, with too little commonality among the elements of the set. Gene therapy, as circumscribed by the NIH Guidelines, includes genuine therapeusis, as well as cell-marking and "cell-trafficking" experiments; experiments intended to replace a genetically deficient molecule, or to overproduce a chemotherapeutic agent; the use of RNA and DNA viruses, or no viruses at all; viruses administered directly to the patient, virus-transformed cells administered to the patient, and the transformation of cells from many different organs. Thus, as "gene therapy" is defined by the NIH, the only common element is that recombinant DNA techniques are used somewhere in conferring a genetic change in something that is administered to the patient. As discussed throughout this volume, scientific consensus holds that this is not a category that makes sense (reviewed in chapter 1).

An inventory needs to be carefully considered if it is to be of value. What specific concerns should it address? The long-term effects of insertional mutagenesis? Of exposure to live viruses? Of potential novel recombinants that might arise from transformed cells?

What scope is appropriate? If live viruses are the concern, why not include patients who have been exposed to live viral vaccines? If the concern is novel recombinants from transformed cells, why not include patients who have had immunotherapy with their own tumor cells modified in some way?

An inventory of gene therapy protocols is a bad idea that won't die. It seems to have originated at the FDA in 1993,[5] and has kicked around since then. In spite of the limitations described above, the concept lives on at NIH. Varmus seems to be using it now as a sop to powerful members of Congress who, without the remotest conception what purpose it serves, demand NIH's continued involvement in the oversight of gene therapy.

If 20 years of struggling with the regulation of biotechnology has taught us anything, it should be that fuzzy thinking and political compromises lead to flawed public policies, while scientific reasoning can light the way. Finally, excessive regulation of human gene therapy—including the NIH's presumptive decision to maintain a vestigial Office of Recombinant DNA Activities, create a new advisory committee, hold conferences and maintain the database—must be viewed in the context of the "reinventing government" downsizing occurring elsewhere. Both the NIH and the FDA, for example, were told to reduce their advisory committees by one-third, regardless of their merits. Many of these advisory committees make a vital contribution: at the FDA, they speed up and lend certainty to the regulatory review of pharmaceuticals, and at the NIH they provide essential peer-review of the arcana of grant proposals and other programs. But Vice President Gore's morbid fascination for the new biotechnology (chapter 2) gains it no end of special attention—and problems.

NON-INNOCENTS ABROAD

The situation is worse abroad. Japanese ministries have been bumping into one another as they craft new, awkward, ponderous mechanisms for regulating gene therapy. The Ministry of Health and Welfare seems inclined toward a system similar to that of the NIH RAC—which will be an anachronism at birth—while the Ministry of Education, Science and Culture, which is responsible for university hospitals, has not made a final decision on its position.[6] As of this writing, only a single patient in Japan had undergone gene therapy.

While no authority comparable to the NIH RAC exists at the level of the European Union (EU), individual countries are moving inexorably toward new requirements for human gene therapy that are at least as burdensome and insupportable as those of the RAC—requirements that certainly exceed those for other innovative, experimental, clinical therapies. The regulatory processes in the United Kingdom and France are the most clearly and formally defined.

In the United Kingdom, over and above its usual mechanisms for overseeing clinical research (the National Health Service and its local research ethics committees, the Medicines Control Agency and so forth), the government has established the Gene Therapy

Advisory Committee (GTAC), whose primary regulatory function is to perform case-by-case review of gene therapy proposals.[7]

In France, in addition to the usual mechanisms for overseeing clinical research, the National Advisory Committee on Ethics (CCNE), the Commission de Gene Genetique (CGG) and the Commission de Genie Biomoleculaire (CGBM) are involved in either policy toward or case-by-case review of gene therapy proposals.[8] Other European countries are at various stages of establishing processes for regulating human gene therapy.

At a time when governments are generally reducing funding for both basic and clinical research, it is disquieting that they are creating gratuitous regulatory disincentives to state-of-the-art scientific investigation. Eliminating unnecessary layers of gene therapy regulation would be a win-win proposition: reduction of unnecessary government and researchers' spending on regulation; more resources available for the actual research and with diminished regulatory disincentives, greater interest in gene therapy from industry.

Gene Therapy for Enhancement

We usually think of "gene therapy" as the insertion of normal or modified genes into somatic (nongerm line) cells of human subjects in order to correct genetic or acquired disorders via the in vivo synthesis of missing, defective or insufficient gene products. But it is likewise possible, in the *absence of disease*, to use the same techniques in order to enhance desirable characteristics, via expression of inserted genes (enhancement engineering). Some of the nuances of the issue of enhancement engineering are illustrated in this statement by W. French Anderson, one of gene therapy's pioneers:

> Our Western culture is very pluralistic and permissive. We, ourselves, might not want to smoke, use krebiozen to treat our cancer, or ride a rocket over the Snake River Canyon, but we allow others to do what they wish with their own lives and bodies, within broad limits, short of suicide or hurting others. Thus, our society may want to allow someday somatic cell genetic engineering by a competent adult for him/herself. But until we have acquired considerable experience with regard to the safety of somatic cell gene therapy for severe disease, and society has resolved at least some of the ethical dilemmas that this procedure

> would produce, nontherapeutic use of genetic engineering should not occur.[9]

Rapid advances in biomedical research stimulate the development of numerous medical technologies and individual products, and their translation into clinical use raises complex—but not necessarily unique—medical, economic, ethical and social issues. The British magazine, *Scrip,* published this item:

> The U.S. biopharmaceutical company, AntiCancer Inc, has developed a DNA delivery system that specifically targets hair follicles. Using an in vitro skin culture system, company researchers said it was possible to determine directly the effects of genes and drugs on hair growth and color. Gene therapy for all types of hair loss, including baldness and chemotherapy-induced alopecia, may be possible, AntiCancer said.[10]

My initial response was incredulous. Gene therapy for baldness seemed like using radiation therapy on warts. But such an application of SHGT, on reflection, might not be so extreme.

Some commentators have suggested that a line can and should be drawn between SHGT for therapeutic or research (cell-marking, for example) purposes on one hand and enhancement engineering on the other.[11] Anderson has argued that although SHGT for the treatment of severe disease is consistent with the moral principle of beneficence—having the potential to alleviate human suffering—the same is not true of enhancement. He cites two additional reasons that enhancement engineering "would threaten important human values:"

First, it could be medically hazardous, i.e., the risk could exceed the potential benefits and could therefore cause harm, and second, it would be morally precarious, i.e., it would require moral decisions that our society is not now prepared to make and which could lead to an increase in inequality and an increase in discriminatory practices.[12]

The nature of the objections on medical grounds is fairly self-evident. Introducing a foreign substance via SHGT is fraught with the same kinds of uncertainty as for a new drug. There are some new wrinkles, to be sure; for example, SHGT might inactivate a cancer suppressor gene or activate an oncogene (but this is similar to concern about carcinogenic or teratogenic potential with more mundane therapies). With SHGT, however, the effect of an unexpected adverse reaction could be compounded by the relative irreversibility of the process.

Anderson's second point, concerning the morality of enhancement, is intriguing. One can certainly describe some scenarios that may verge on "moral precariousness:" the introduction into a patient of a gene for an appetite-suppressant hormone, a gene coding for a brain chemical that enhances memory, an additional growth hormone gene in a hormonally normal but shorter-than-average adolescent who has short parents, an additional growth hormone gene for an adolescent of normal height who wishes to play professional basketball, and a gene conferring resistance to industrial toxins for someone who works with hazardous chemicals. These scenarios raise issues that are pertinent, but by no means unique, to SHGT for enhancement:

- *The difficulty of defining whether a patient's disease is serious* (or, in fact, whether there is a disease at all). One might consider obesity a serious disease at triple the ideal weight, whereas a person 30% above ideal weight might be considered to have a minor disease, and someone near ideal weight but wishing to be thinner for a trip to the Caribbean could be said to suffer only from cultural discomfort. Physicians currently make such judgments, routinely in order to determine whether various interventions—special diets, chemical appetite suppressants or surgical operations to induce malabsorption of nutrients—should be prescribed.
- *Equality of access to SHGT.* Which societal models will be invoked to determine who gets the therapy? Those with the greatest or most acute need? Those best able to benefit society? First come/first served? Ability to pay? We already have models for allocating various therapies ranging from renal and hepatic transplants to rhinoplasties.
- *The possibility of coercion.* Would pressure be exerted on workers to have gene therapy to make them less susceptible to workplace toxins? It was not so long ago, after all, that women working with CRT's were required by employers to promise not to become pregnant.

- *Gene therapy as a therapy for discrimination.* Would people of color seek genetic "improvement" of skin color or other traits, in order to obtain greater economic and social opportunities for themselves or their progeny?

Anderson concluded that stepping over the line that separates therapy from enhancement would be opening Pandora's box, and that on both medical and ethical grounds, any form of enhancement engineering should be excluded.

The questions that have been raised by Anderson, bioethicist John Fletcher and others are important ones, and their approach to them has been scholarly. However, their arguments largely ignore the indications for which the testing of medical therapies already occurs. They do not give adequate weight to the latitude that our society affords to citizens who wish to enhance their physical appearance or their health. More prosaic therapies, commonly rubbed on or swallowed, are often intended, after all, only to grow hair, decrease appetite or lighten age spots. On what ethical basis should SHGT be subjected to a higher standard than liposuction, radial keratotomy or the application of permanent makeup by tattooing?

SHGT is part of a therapeutic continuum that includes allogeneic organ transplantation, the injection of live viruses for vaccination, and the administration of drugs to activate dormant genes.[13] The medical and ethical issues raised by SHGT, however difficult to resolve they may be, are hardly unique. There is no obvious basis for invoking a discriminatory, more restrictive paradigm for SHGT than for analogous experimental therapies. Arguments against prescribing SHGT for enhancement should be weighed against society's permissiveness toward experimental medical interventions generally, and those intended for enhancement in particular.

First, consider Anderson's argument that SHGT may be medically hazardous and irreversible. This objection can be addressed by regulators requiring that, when appropriate, experiments be carried out in a way that is not irreversible: inserting the introduced gene that is to be expressed in a way that it is inducible and under positive control; transducing only cells, such as lymphocytes, that have a finite lifespan or can be sequestered; or killed or making the transduced cells surgically accessible (for example, in a skin graft), for ease of removal. These techniques have already been accomplished or are technically feasible.

Second, and philosophically more important, is an argument by analogy. For better or worse, drugs are not infrequently tested for relatively trivial indications, such as modest obesity, stuffy nose and baldness. There have been numerous clinical trials of appetite suppressants, memory-enhancing or mentation-improving drugs, and human growth hormone for hormonally normal but shorter-than-average children. The large array of entities, both local and federal, that regulate SHGT—Institutional Review Boards (IRBs), Institutional Biosafety Committees (IBCs), the NIH RAC, the NIH director and the FDA—contrasts sharply with the degree of oversight of, say, a new surgical procedure, which might be completely unregulated or subject only to approval of a hospital-based committee; or tattooing, which is often overseen only by a municipal licensing authority.

One should also consider the wishes and well-being of patients. Enhancement is not invariably regarded as trivial—certainly not by the adolescent boy who is six inches shorter than anyone else in his class, or to many people of either sex who suffer hair loss. One need look no further than the huge societal demand for cosmetics, cosmetic surgery and health clubs, to be reminded how important people consider it is to look and feel good.

The issues surrounding whether a patient suffers from a serious disease (or any disease), equal access to therapy, the possibility of coercion, and the relationship between medical intervention and discrimination, are fundamentally no different for SHGT than for other medical interventions. Therefore, innovations such as SHGT, even when used for enhancement, should arguably be treated similarly to other analogous medical interventions, except as scientific considerations may dictate. Such innovations should certainly not be rejected out of hand, particularly when existing societal oversight mechanisms subject them to a high level of case-by-case review.

Support for a rational approach to SHGT for enhancement has come from disparate sources. Georgetown University bioethicist LeRoy Walters has argued that we should consider the momentous positive impacts on both individuals and society that somatic cell enhancement engineering could have.[14] With commendable libertarian zeal, *The Economist* editorialized,

> what of genes that might make a good body better, rather than make a bad one good? Should people be able to retrofit themselves with extra neurotransmitters to enhance various mental powers? Or to change the color of their skin? Or to help them run faster, or lift heavier weights? Yes, they should. Within some limits, people have a right to make what they want of their lives.[15]

Some have asserted that any form of enhancement engineering should be proscribed. However, several arguments may be marshaled against that view: medical risks of SHGT appear not to be materially different from those of other kinds of medical interventions and are likely to be controllable, or at least evaluable; medical interventions for enhancement are often regarded by patients as not at all trivial; close analogies exist in other highly-regulated as well as minimally-regulated medical interventions; and existing societal oversight mechanisms are more than adequate to balance the risks and benefits of proposed clinical protocols.

If society is to realize the full spectrum of benefits from human gene therapy, it cannot be considered in a philosophical vacuum from which relevant precedents and experience are excluded. It must be judged in the broader context of what people want and what society permits.

REFERENCES

Introduction

1. Anon. Coordinated Framework for Regulation of Biotechnology. Federal Register 1986; 51:23302-23347.
2. Idem.
3. Idem.
4. U.S. Department of Agriculture Animal and Plant Health Inspection Service. Plant pests; introduction of genetically engineered organisms or products. Federal Register 1987; 52:22892-22915.

Essay 1: FDA

1. Laffer WG, Bord NA. George Bush's Hidden Tax: The Explosion in Regulation. The Heritage Foundation Backgrounder; Washington D.C.: July 10, 1992.
2. The Boston Consulting Group analysis, as quoted by the Office of Technology Assessment in Pharmaceutical R&D: Costs, Risks, and Rewards, Washington D.C., February 1993. (Estimate is in pretax 1990 dollars.)

3. Piercey L. New gender analysis rules could impact drug development. BioWorld Today 1995; 6:1-5.
4. Miller HI. Anti-medicine man. National Review 1995; 48-51.
5. Henderson DR. FDA censorship can be hazardous to your health. Policy Brief 158, St. Louis, Mo: Center for the Study of American Business; September 1995.
6. Anonymous. Adverse Experience Reporting Requirements for Human Drug and Licensed Biological Products. Federal Register 1995; 59.
7. Meinert CL. Controlled Clinical Trials; (in press).
8. Henderson, Ibid.
9. Henderson, Ibid.
10. Henderson, Ibid.
11. Henderson, Ibid.
12. Friedman M. Why government is the problem. Stanford: Hoover Institution Essays in Public Policy 1993:8-9.
13. Schwartz J. FDA revises biotechnology rules. Washington Post 1995; A29.
14. Anon. Coordinated Framework, Ibid.
15. Sternberg S. Three days of debate yields possible consensus on FDA definition for biotech products. BioWorld 1995; 6:1-4.
16. FDA press release, March 29, 1996.
17. Anon. Improving the nation's drug approval process. Council on Competitiveness Fact Sheet, November 1991.
18. Anon. The Impact of the House FDA Reform Proposals. Food and Drug Administration, 1996.
19. Giaquinto AR. Letter to House Commerce Committee from Schering Plough Resarch Institute. May 14, 1996.
20. Anon. A National Survey of Oncologists Regarding the Food and Drug Administration. Washington D.C.: The Competitive Enterprise Institute 1995; 1-4.
21. Huttner SL. Biotechnology and Food. New York: American Council on Science and Health, 1996.
22. Anon. Foods Derived from New Plant Varieties: Consultation Procedures. Center for Food Safety and Applied Nutrition and Center for Veterinary Medicine, Food and Drug Administration. June 1996.
23. Anon. Statement of Policy: Foods Derived From New Plant Varieties. Federal Register 1992; 57:22984-23005.
24. Nordler JA, Taylor SL, Townsend JA et al. Identification of a Brazil nut allergen in transgenic soybeans. New Eng J Med 1996; 334:688-699.
25. Taylor VJ. Memorandum to the Labeling Subcommittee, California Interagency Biotechnology Task Force, April 15, 1994.

Essay 2: EPA

1. Lincoln DR, Fisher ES, Lambert D et al. Release and containment of microorganisms from applied genetics activities. Report prepared under EPA Grant #R-808317-01, 1983.

2. Seidler RJ, Hern S. Special report: the release of ice-minus recombinant bacteria at California test sites. Environmental Protection Agency, Environmental Research Laboratory, Corvallis, Oregon, March 1988.
3. Idem.
4. Bragg JR et al. Effectiveness of bioremediation for the Exxon Valdez oil spill. Nature 1994; 368:413-418.
5. Reilly W. Text of speech at Stanford University, undated.
6. Miller HI. A Need to Reinvent Biotechnology Regulation at the EPA. Science 1994:266, 1815 and vide infra.
7. House Appropriations Committee staff, personal communication.
8. Anon. Microbial Pesticides: Experimental Use Permits and Notifications. Federal Register 1994; 59:45600. See also, Report of the Joint EPA FIFRA Scientific Advisory Panel and Biotechnology Science Advisory Committee Subpanel on Plant Pesticides, February 10, 1994.
9. Miller HI. Concepts of risk assessment: the "process versus product" controversy put to rest. In: Brauer D, ed. Biotechnology. Weinheim: VCH, 1995.
10. Miller. A need to reinvent biotechnology regulation at the EPA. Ibid. See also B.W. Falk and G. Bruening, Science 1994; 263:1395; and Anon.1992 National Biotechnology Policy Board Report, National Institutes of Health, Office of the Director, Bethesda, Maryland, 1992.
11. Anon. Biosafety in microbiological and biomedical laboratories. Centers for Disease Control and National Institutes of Health, U.S. Department of Health and Human Services, U.S. Government Printing Office, Washington, 1988. See also Anon. Field Testing Genetically Modified Organisms: Framework for Decisions, Washington D.C.: National Academy Press 1989 and Miller HI, Burris RH, Vidaver AK et al. Risk-based oversight of experiments in the environment. Science 1990; 250:490.
12. Fumento M. Science Under Siege, New York: Morrow, 1993:19-44.
13. EPA press release, February 1, 1989; and EPA targets chemical used on apples. Washington Post, February 2 1989:4A(news item).
14. Shabecoff P. Hazard reported in apple chemical: E.P.A. cites a risk of cancer but will not bar use yet. New York Times, February 2, 1989:23.
15. Fumento, Ibid. See also Bidinotto RJ. The great apple scare. Reader's Digest 1990; 137:5556.
16. Federal Register 59. Ibid.
17. Anon. Field Testing Genetically Modified Organisms: Framework for Decisions. Ibid.
18. National Biotechnology Policy Board Report, Bethesda MD: National Institutes of Health, Office of the Director,1992. See also Federal Register 1992; 57:6753.
19. Miller HI. Concepts of Risk Assessment. Ibid.

20. Anon. Final Rule on the Testing of Microbial Pesticides. EPA background document accompanying the biotechnology microbial pesticides rule. EPA Office of Prevention, Pesticides and Toxic Substances, August 1994. See also Federal Register 59, Ibid.
21. Anon. Coordinated Framework for Regulation of Biotechnology. Federal Register 1986; 51:23302-23347.
22. Anon. Report of the Joint EPA FIFRA Scientific Advisory Panel and Biotechnology Science Advisory Committee Subpanel on Plant Pesticides. Ibid. See also Anon. Appropriate Oversight for Plants with Inherited Traits for Resistance to Pests, 1996.
23. Anon. Field Testing Genetically Modified Organisms: Framework for Decisions. Ibid.
24. Goodman RM, Hauptli H, Crossway A et al, Science 1987; 236:48.
25. Idem.
26. Moffatt AS, Science 1994; 265:1804.
27. Anon. Report of Joint EPA FIFRA Panel. Ibid.
28. Huttner SL. Government, researchers, and activists: the critical public policy interface. In: Brauer D, ed. Biotechnology. Weinheim: VCH, 1995.
29. Anon. Field Testing: Framework for Decisions. Ibid.
30. Anon. Safeguarding the Future: Credible Science, Credible Decisions, The Report of the Expert Panel on the Role of Science at EPA. EPA Document 600/9-91/050, March 1992.
31. Hoyle R. Bio/Technology 1992; 10:742.
32. Fisher LJ. EPA Assistant Administrator letter to Dennis Focht, August 21, 1992.
33. Stone R. EPA gives science adviser more clout. Science 1995; 267:1895.
34. Fox JL. EPA's first commercial release is still pending. Bio/Technology 1995; 13:115.
35. Cook RJ, Letter dated September 17, 1992, to the Environmental Protection Agency Public Response Section, Field Operations Division.
36. Hoskin AF, Leigh JP, and Planek TW. Estimated risk of occupational fatalities associated with hazardous waste site remediation. Risk Analysis 1994; 14:1011-1017.
37. Breyer S. Breaking the Vicious Circle. Cambridge: Harvard Press, 1993:23.
38. Reilly, W. Speech at Stanford University, January 1994.
39. Taylor J. Salting the earth. Regulation 1995; 2:53-66.
40. Wildavsky A. Wealthier is healthier. Regulation 1980; 4:10.
41. Keeney RL. Mortality risks induced by economic expenditures. Risk Analysis 1990; 147:148. See also Catalano R. The health effects of economic insecurity. Am J Public Health 1991; 81:1148.

Reference for Box

1. Anon. Appropriate Oversight for Plants with Inherited Traits for Resistance to Pests, 1996.

Essay 3: USDA

1. U.S. Department of Agriculture Animal and Plant Health Inspection Service. Plant pests; introduction of genetically engineered organisms or products. Federal Register 1987; 52:22892-22915. Also, U.S. Department of Agriculture Animal and Plant Health Inspection Service. Genetically engineered organisms and products; Notification procedures for the introduction of certain regulated articles; and Petition for nonregulated status Federal Register 1993; 58:17044-17059.
2. Medley TL, McCammon SL. Strategic regulations for safe development of transgenic plants. In: Brauer D, ed. BIOTECHNOLOGY. Weinheim: VCH, 1995:197-212.
3. Huttner, SL. Government, researchers and activists: the crucial public policy interface. In: Brauer D, ed. BIOTECHNOLOGY. Weinheim: VCH, 1995:459-494. Also, Arntzen CJ. Regulation of transgenic plants. Science 1992; 257:1327; and Huttner SL, Arntzen C, Beachy R et al. Revising oversight of genetically modified plants, Bio/Technology 1992; 10:967-71; and Huttner, SL. Risk and reason: an assessment of APHIS. In: Weaver RD, ed. U.S. Agricultural Research: Strategic Challenges and Options. Bethesda: Agricultural Research Institute, 1993.
4. Moses PB, Hess CE. Getting biotech into the field. Issues in Science and Technology, Fall 1987:35-40.
5. Moses and Hess, Ibid.
6. Anon. Field Testing Genetically Modified Organisms: Framework for Decisions. Washington D.C.: National Academy Press 1989.
7. Coleman M. Personal communication.
8. Miller HI, Burris RH, Vidaver AK et al. Risk-based oversight of experiments in the environment. Science 1990; 250:490-92.
9. Anon. Exercise of federal oversight within scope of statutory authority: planned introductions of biotechnology products into the environment. Federal Register 1992; 57:6753-62.
10. Miller HI, Regulation. In: BD Davis, ed. The Genetic Revolution. Baltimore: Johns Hopkins University Press, 1991:196-211.
11. Anon. Exercise of federal oversight within scope of statutory authority. Ibid.
12. Rabino I. The Impact of Activist Pressures on Recombinant DNA Research. Science Technology and Human Values 1991; 16:70-87.
13. Ratner M. Survey and Opinions: Barriers to Field-Testing Genetically Modified Organisms. Bio/Technology 1990; 8:196-198.
14. Sears R. Letter on Behalf of the American Society of Horticultural Sciences, American Phytopathology Society, and Crop Science Society of American to the EPA. Docket control number OPP-300370, February 4, 1995.
15. Anon. Exercise of federal oversight within scope of statutory authority. Ibid.

16. Anon. Guidelines Recommended to USDA by the ABRAC, March 1992, USDA Document No. 91-04.
17. Kaiser J. USDA kills biotech office, advisory panel. Science 1995; 270:1911.
18. Covello VT, Fiksel JR. The Suitability and Applicability of Risk Assessment Methods for Environmental Applications of Biotechnology, Report No. NSF/PRA 8502286, Washington D.C.: National Science Foundation, 1985.

Essay 4: NIH

1. Berg P, Singer MF. The recombinant DNA controversy: 20 years later. Proc Natl Acad Sci 1995; 92:9011-9013.
2. NIH and FDA press releases; for background, see Anon. FDA and NIH will consolidate gene transfer protocol review by December. F-D-C Reports, July 25, 1994:13-14.
3. Anon. Recombinant DNA Research: Notice of Intent to Propose Amendments to the NIH Guidelines for Research Involving Recombinant DNA Molecules (NIH Guidelines) Regarding Enhanced Mechanisms for NIH Oversight of Recombinant DNA Activities. Federal Register 1996; 61:35774-35777. See also Anon. Varmus proposes to scrap the RAC. Science 1996; 272:945.
4. Idem.
5. Anonymous. Gene therapy patient registry for long-term side effects proposed by FDA. F-D-C Reports. November 1, 1993:19.
6. Barker S. Guidelines may boost gene therapy in Japan. Nature 1995; 378:327.
7. Macer DRJ, Akiyama S, Alora AT et al. International perceptions and approval of gene therapy. Human Gene Therapy 1995; 6:791-803.
8. Idem.
9. Anderson WF. Human gene therapy: why draw a line? Med Philos 1989; 14:681-90.
10. Anon. Gene therapy for hair loss. Scrip 1993:27.
11. Anderson WF. Prospects for human gene therapy. Science 1984; 226:401-09. And Anderson WF, Fletcher JC. Gene therapy in human beings: when is it ethical to begin? New Engl J Med 1980; 301:1293-97.
12. Anderson. Human gene therapy: why draw the line. Ibid.
13. Miller HI. Human gene therapy: part of a therapeutic continuum. Hum Gene Ther 1990; 1:3-4.
14. Walters L. Presentation at Genetics, Religion, and Ethics Conference Houston, Texas, March 16, 1992.
15. Anon. Changing your genes. The Economist 1992:11-12.

CHAPTER 4

Ripple Effects in the Real World

ESSAY 1
REAL-WORLD EFFECTS OF FLAWED REGULATION

Biotechnology offers striking illustrations of science policy built on a foundation of invalid scientific assumptions, pseudocontroversy and political or ideological goals. One result is expensive, expansive and irrational regulation. The arguments marshaled during the biotechnology policy "debate" are revealing. Those who would encourage unnecessary regulation sometimes argue that, in the face of uncertainty, it is only prudent to "err on the side of safety," to avoid taking any chances, and act instead on the basis of the worst-case scenario. A related argument is that even if only a handful of cancers or poisonings (or genetically engineered tomato plants running amok) are prevented by government oversight regimens, not to act would be unconscionable and would amount to putting a price on human life.

But the principles of "erring on the side of safety" and the pricelessness of life do not withstand rigorous scrutiny. What appears to be the "safe" choice may actually pose greater risk. What is required, then, is a "comparative risk assessment" that considers the risks of various alternative courses of action. Consider, for example, this analysis by Supreme Court Justice [then federal Court of Appeals Chief Judge] Stephen Breyer:

> the regulation of small risks can produce inconsistent results, for it can cause more harm to health than it prevents. Sometimes risk estimates leave out important countervailing lethal effects, such as the effect of floating

> asbestos fibers on passersby or on asbestos-removal workers... Sometimes the regulator does not, or cannot easily, take account of offsetting consumer behavior, as, for example, when a farmer, deprived of his small-cancer-risk artificial pesticide, grows a new, hardier crop variety that contains more "natural pesticides" which may be equally or more carcinogenic.[1]

An important proposition related to the effort and expense of regulatory compliance pertains to the direct correlation between wealth and health, an issue popularized by the late political scientist Aaron Wildavsky. It is no coincidence, the argument goes, that richer societies have lower mortality rates than poorer ones, and to deprive members of society of wealth is to enhance their risks.[2] Wealthier individuals obtain better health care, more nutritious diets and generally less stressful lives; conversely, the deprivation of income itself has adverse health effects, in the form of poorer diet, restricted access to health care and more stress. These lead, in turn, to a disproportionate impact of "treatable" disease in poor populations and an increase in stress-related problems, including ulcers, hypertension, heart attacks, depression, accidents and suicides. The difference can be characterized as a "risk gap" between the poor and wealthier segments of society.

It is difficult to quantify the relationship between the deprivation of income and mortality, but a sampling of academic studies suggests, as a conservative estimate, that every $7.25 million of regulatory costs *induces* one additional fatality through this "income effect."[3]

There are direct and indirect costs of the government's regulatory policies, and the effects are the same. When the government's own costs for implementing unneeded regulations raise the federal budget, there is a need for additional revenues from taxes. In this way, diverting discretionary income from taxpayers to its own coffers government directly elicits the income effect. Government's actions exert the same effect indirectly, as overly cautious and expensive regulatory requirements inflate research and development costs for useful new products. The increased costs are passed along to consumers, which again elicits the income effect.

What I term "comparative risk management" is the critical decision-making process that determines how we make choices where to spend the regulatory dollar. Jerry Taylor of the Cato Institute argues that it may best not be spent on regulation at all:

> Money spent on Superfund risks is money not spent on something else, including the ability to protect public health in other ways, reduce poverty, improve public safety, or even the intangible (but very real) benefits one gets from disposing one's income as one likes. According to the EPA's (extremely dubious) estimates, for example, $1 million spent on Superfund saves approximately 2.5 years of life. But $1 million spent on breast cancer screening saves 300-700 years of life. Similarly, $1 million spent on cervical cancer screening saves 700-1500 years of life.[4]

Most of the examples in the preceding chapters pertain to the United States. Internationally, regulation has been a hodgepodge, and often excessive. No nation has demonstrated a monopoly either on government officials' virtuosity or villainy. As a showcase for many useful examples, it is worth considering the international stage on which biotechnology regulatory debates are being played out.

If there are generalizations to be made, it appears that the farther removed an environmental regulatory organization is from the taxpayers, the less scientific, the more political and self-serving are its actions. As that axiom would predict, the supranational organizations such as the United Nations and the European Union have been among the worst (*vide infra*).

EUROPE

In much of Europe, where oversight has focused on "GMOs" (genetically modified organisms, narrowly defined as those manipulated with recombinant DNA techniques), regulatory burdens have had significant impact.[5] Denmark prohibits the patenting or use of animals that have had genetic functions added or deleted with the newest techniques. The European Union's (EU) directives 90/219/EEC and 90/220/EEC are similarly focused on recombinant DNA techniques.[6] The European Union (EU) and certain European nations, especially Germany and Denmark, have created potent regulatory disincentives to the use of new biotechnology.[7]

To use a laboratory metaphor, Germany might be considered to be the "positive control" experiment in which regressive regulation is combined with regional antiscience activism. It turns out to be a volatile mixture.

Burdened by both the European Union's and national regulations, German researchers have found themselves in a regulatory stranglehold. They are hindered by required "case-by-case-every-case" governmental review, even of negligible-risk experiments. They have low expectations that products ultimately will ever be approved for marketing and are beleaguered by activists who manifest a degree of hostility not seen elsewhere. Consequently, many scientists and companies have left the country and some who remain are conducting field trials abroad.[8]

Of some 6000 field trials world wide of plants genetically engineered with the most precise gene-splicing techniques, only a few dozen have been in Germany. It is particularly disturbing that in 1995, all 15 of such small-scale field trials conducted by universities and research institutes in Germany were partially or completely destroyed by activists, even though most of the experiments were studying the environmental safety of growing genetically manipulated plants in normal agricultural environments. One postdoctoral fellow was attacked with stones while trying to protect his virus-resistant sugar beets from vandals.[9]

The current situation recalls Germany in the 1930s when the Third Reich vilified and persecuted the practitioners of what the regime called *Entartete Kunst*, "degenerate art." The artists, accused by propaganda minister Joseph Goebbels of "insolent arrogance" and "snobbism," included such dangerous subversives as Emil Nolde, Max Beckmann, Marc Chagall, Vincent van Gogh, Henri Matisse, Edvard Munch and Pablo Picasso. Now we have a kind of *Entartete Forschung*, "degenerate research." The stridency and absolutism of the activists' pronouncements—and their violent tendencies—are uncomfortably familiar. The German government is not culpable in the current situation, however, except indirectly by neglecting to protect the personal safety and property of scientists against assaults by antibiotechnology activists. But the vandals *are* abetted by governmental ambivalence and policies that equate innovation with risk. Flying in the face of the scientific consensus, current EU and German regulation casts a veil of suspicion over biotechnology by requiring case-by-case government environmental assessments for field testing with rDNA-manipulated plants. By contrast, plants with similar or even identical traits that were created with less precise techniques, such as hybridization or mutagenesis, are subject to no government scrutiny or requirements (and no publicity) at all. And that applies even to the numerous new plant varieties that result from "wide crosses," hybridizations which move genes from one species or genus to another—that is, across natural breeding boundaries. If rDNA-manipulated plants were treated appropriately—that is, like other new varieties—their testing would not need government review, special warning signs or public announcements. There would be no way for the thugs to target and disrupt field research that they have decided are *Entartete Forschung.*

The German experience is an important corollary of the law of unintended consequences: the problem would have been avoided entirely, had public policy been crafted intelligently in the first place.

Five years after the EU's regressive, process-based directives (the equivalent of regulations in the United States) were promulgated, EU countries continue to struggle with their implementation. The directives, specific for GMOs, encompass both contained uses and planned introductions (that is, field trials and commercial uses).[10]

In 1990, the EU's Directorate General XI (Environment) rushed to publish the process-based directives, in order to preempt reports from the scientifically stronger Group of National Experts on Biotechnology of the OECD (Organization for Economic Cooperation and Development, based in Paris). (The various OECD reports reiterated that neither new regulatory paradigms nor new regulations were necessary for biotechnology products.) Not surprisingly, the directives have dramatically hindered research and development in member countries.[11] They have the potential to affect other countries, as well, through thinly veiled provisions that can be used to erect nontariff trade barriers to foreign rDNA-derived products.

The United States Congress' Office of Technology Assessment published this analysis in 1991:

> In enacting directives that specifically regulate genetically modified organisms, the EC [now the EU] has established a regulatory procedure that is significantly different from that of the United States. In the EC, regulation is explicitly based on the method by which the organism has been produced, rather than on the intended use of the product.

> This implies that the products of biotechnology are inherently risky, a view that has been rejected by regulatory authorities in the United States. In addition, manufacturers are concerned that their new biotechnology-derived products may face additional barriers before they can be marketed, for the product may also be subject to further regulations based on its intended use.[12]

The authoritative October 1993 report of the UK House of Lords Select Committee on Science and Technology, *Regulation of the United Kingdom Biotechnology Industry and Global Competitiveness,* echoed other assessments of the EU's approach to biotechnology. It excoriated the EU's regulatory approach to both contained uses and field trials of organisms and recommended reduced and rationalized regulation:

> As a matter of principle, GMO-derived products should be regulated according to the same criteria as any other product... U.K. regulation of the new biotechnology of genetic modification is excessively precautionary, obsolescent and unscientific. The resulting bureaucracy, cost and delay impose an unnecessary burden to academic researchers and industry alike.[13]

As George Orwell said, to see what is in front of one's nose requires a constant struggle—a struggle that the European envirocrats have been losing. The primary goal of the EU's environmental regulators seems to be the maintenance of a large, centralized and unnecessary bureaucracy. The regulators occasionally winnow out certain field trials from case by case review, a process they tout as being "risk based," but the basic paradigm is flawed. And bad science makes for bad regulation.

A blunder of huge proportions typical of the EU's approach to biotechnology is the policy on bST (bovine somatotropin or bovine growth hormone), a protein hormone that increases dairy cows' productivity. In December 1994, the EU Agricultural Council voted to extend its moratorium on commercial use of bST until December 31, 1999.[14] However, the EU permits the importation of dairy products from countries such as the United States which use the product extensively (approximately 15% of dairy cows in the U.S. are treated with bST)—and whose dairy farmers, therefore, benefit from enhanced productivity. Thus, EU policy makers have created what is, in effect, a *reverse* nontariff trade barrier: that is, they've disadvantaged European dairy products in their own markets!

Policy aside, the EU has also made unwise and anti-innovative decisions on individual products. In the Spring of 1996, an EU regulatory body consisting of representatives from member states and established under Directive 90/220 on Deliberate Release of Genetically Modified Organisms into the Environment denied approval of Ciba-Geigy's recombinant DNA-derived pesticide-resistant maize seed.[15] The regulators offered two reasons: the seed was considered not to be sufficiently labeled, and it would poison insect larvae. As discussed in chapter 3, however, there is no justification for product labeling to disclose the genetic technique(s) used to construct a plant variety. As to the second point, one wonders why European regulators seem not to be equally concerned about the pesticidal properties commonly found in the plants that humans routinely consume, or present in widely used "natural" microbial biocontrol agents such as *Baculovirus* and *Bacillus thuringiensis*.

The rejection by the regulatory committee will be passed onto the Council of Ministers who can choose to accept the rejection or favor approval; in the latter case, the matter would be sent back to the regulatory body for input from the executive body.[16]

European parliamentarians and regulators are beginning to get a taste of the bitter fruits of their own policies and product decisions. Several hundred people attended a January 1996 get-together in Brussels, an informal conclave of the European Commission, European Council and European Parliament devoted exclusively to biotechnology. The data presented were ominous for Europe, indicating that the United States enjoys a substantive lead over all of Europe combined, in virtually every category, including number of biotechnology companies (1300 versus 485), biotechnology patents (65% versus 15%), and biotech R&D expenditure (7 billion ECU versus 2.2 billion ECU). The European officials concluded, according to *BioWorld*, that "unless member countries quickly develop the economic, legal and ethical framework to foster biotechnology business development, the continent is in danger of becoming merely a market rather than a major contributor to biopharmaceutical innovation."[17]

But the attendees were unable to make the connection between their regressive and stultifying regulatory policies and diminished R&D. In the end, they reverted to type, calling for more public discussion of ethical issues supposedly raised by biotechnology, in particular "vague fears" that the "value of the human being, founded on its genetic constitution, may no longer be established" (whatever that might mean). Such dithering accomplishes nothing. The anthem for the European Union is Beethoven's sublime "Ode to Joy," but a more appropriate choice would be the musical theme from the British political parody, "Yes, Minister."

JAPAN

Although Japan's regulatory approach has been illogical and process-based, restrictions on pharmaceutical products made with the newest techniques have not been debilitating to industry, and the new biotechnology applied to pharmaceuticals there has sustained significant commercial progress. But some specific areas undoubtedly have been impeded by the government's conviction that the use of rDNA techniques, per se, raises new safety issues. For example, despite a medical and scientific infrastructure that could support substantial clinical trials of human gene therapy, only one patient has been treated in Japan's only gene therapy trial, and not a single company has been created there with gene therapy as its goal. By contrast, gene therapy trials are already well under way in the United States, Italy, France, the Netherlands, England, Germany, Poland and China. More than 600 patients have been treated and the numbers are rising rapidly.

Japan's stigmatization of the new biotechnology is similarly reflected in the dearth of activity in agricultural biotechnology. Only about a half-dozen Japanese field trials of recombinant DNA-manipulated plants (and none of microorganisms) have been carried out (of approximately 6000 performed world-wide), and Japanese research and development in this area is far behind what one would expect. The Japanese government has provided little encouragement in the form of clear, predictable, risk-based regulations to those contemplating field trials. The Japanese Ministry of Health and Welfare has imposed a strict regulatory regime specific to foods and food additives manufactured with rDNA techniques.[18]

With the exception of biotechnology applied to the development of pharmaceuticals (and gene therapy an exception to the exception), Japan is regulating itself out of its rightful share of the biotechnology revolution.

The Developing World

The United Nations and World Bank "median" estimates for the world population in the year 2010 converge at 11.3 billion people, more than double the present 5.5 billion.[19] These estimates are based on the assumption that total fertility rates (TFRs) everywhere will reach replacement value (TFR = 2.06) early in the 21st century. The population estimates are highly sensitive to this assumption; for example, if the long-run TFR is 2.17 instead of 2.06, the population in 2100 becomes *17 billion* instead of 11.3 billion.[20] Over 80% of the added people will live in developing countries.

To feed a population this size will require enormous increases in agricultural production, with special emphasis on production in regions of the most rapid population growth. The past record is, at first glance, encouraging: From 1960-1990, food production in developing countries outpaced population growth.[21] Increased food production has raised caloric consumption rates steadily to an average of about 2500 calories per day[22] and has provided many developing countries with the means to create and sustain a higher degree of economic development. In the 21st century, the new biotechnology may prove to be a watershed for production systems. Recombinant DNA techniques, in particular, are being applied directly and indirectly to a large number of agricultural problems, including the improvement of pest and weed management, plant agronomic properties and postharvest qualities.[23]

The likelihood of future crises in food production, however, demands broader improvements across the spectrum of major local staple crops. Moreover, innovations that increase agricultural productivity while decreasing reliance on agricultural chemicals and other inputs can bring substantial environmental benefits. In many ways, agricultural biotechnology's greatest potential lies in developing countries. Yet, the regulatory approach being pursued by these same nations—on the strong encouragement of industrialized nations—is likely to undermine their chances of efficiently applying the new agricultural tools to their subsistence and export needs. Direct and indirect regulatory costs present tangible and potent disincentives to research, development and product introduction—disincentives that can neutralize all manner of advantages and resources.

Just as in industrialized nations, developing countries' activists' concerns about environmental damage threaten to stall critical biotechnology research and development. Some groups and individuals have been skeptical of and antagonistic toward even the testing of the products of new technologies. Most often their ostensible concerns are about environmental safety. What they neglect to address is that as the world's population grows, the demand for food and the concomitant need to bring more of the world's land mass into production will likely pose a greater threat to the environment than any other single factor. In this conundrum, we see again the need for comparative risk assessment.

In the 1980s per capita yields of the major cereal grains leveled off and then declined on a global basis (although in some of the areas subject to the most rapid population growth it continued to increase).[24] There are whole regions in which food security is severely threatened. On a global basis hunger remains a serious problem for perhaps as many as a billion people. There is a consensus among agricultural scientists that the required increases in food production—amounting to a doubling by the year 2010 in order to feed the number of people projected in the UN median estimate—will have to take place through intensification of production on an essentially fixed land base.

Smil[25] has examined the potential sources for such improvement and estimates that various efficiency gains—via agronomic practices, water management, and reduced waste—along with dietary changes could accomplish gains of about 60%. He accepts that there will be concomitant "worrisome changes" represented by exhaustion of nonrenewable water supplies, soil erosion, salinization and loss of biodiversity, but asserts that these can be prevented or reversed by rational agronomic practices.[26]

The gap remaining is equivalent to not feeding 3 billion people. Smil believes it can be closed by intensifying inputs of fertilizer and water. Leaving aside the economic barriers to supplying these inputs, the margin of safety in optimistic analyses like Smil's is dangerously slim. That margin disappears if the population projections are only a little too optimistic, leaving regions of rapid population growth desperately vulnerable to food security crises.

What happens when Smil's production-intensification strategies fail to meet the growing food demands? Historically, agricultural activity is forced into less arable yet accessible land, converting native ecosystems into limited-production cropland. Intensification failure is further magnified as the environmental deterioration resulting from "up-slope" agricultural movement, in turn, intensifies the problems of rural households in meeting other subsistence needs, such as obtaining firewood and water. As the most basic needs become more labor intensive, the value of children as household assets increases, creating a "vicious cycle" of environmental deterioration and population growth.[27]

Threats to biological diversity warrant special attention. Loss of biological diversity—the one consequence of human population growth that is not reversible—has three primary causes: land use, land cover change and the extinction of domestic species resulting from the introduction of exotic organisms. Among these, the increased pressure on natural ecosystems exerted by agricultural expansion is likely to be the predominant modulator of biodiversity in the 21st century, made better or worse by governmental policies on R&D and their regulation.

Given the narrow margin between productive capacity and the world's projected population, avoidance of these potentially cataclysmic outcomes is likely to depend on *all* the means that can be found to enhance agricultural productivity. And among those means, genetic manipulation of crops, their pests and the enemies of their pests are likely to be especially significant. Regulatory systems that provide adequate protection for ecosystems without imposing unnecessary burdens on the introduction of beneficial changes or technologies will play a pivotal role in determining the positive impact of these new technologies.

Technological advances in farm productivity with concomitant environmental benefits are clearly illustrated by India's recent history with one crop and the Green Revolution. During the period 1961-1966, Indian farmers required 13 million hectares to achieve wheat production levels of 0.83 tons per hectare. With the simultaneous adoption of genetically improved varieties and intensive crop management practices (including pest control and fertilization), production increased five-fold. Since the mid-1970s, the amount of land devoted to wheat production has leveled off at just over 20 million hectares. In the absence of these innovations, by 1991 an additional *42 million hectares* would have been needed to achieve the same production level.[28]

Yet, current trends in biotechnology regulation (see following essay) would limit the development and utilization of these same kinds of genetic improvements when—and only when—plant breeders use the most sophisticated new genetic techniques. Such regulation would threaten biodiversity both by thwarting production intensification needed to reduce conversion of natural ecosystems into cropland, and by diverting regulatory resources away from management of the much more serious threats posed by introductions of disruptive exotic organisms.

The scheme is simply backwards. Regulation has stalled innovation but left the environment unprotected. The organisms that conceivably pose the *greatest* threat of ecosystem disruption and reduction of biodiversity receive the *least* regulatory attention (and vice versa). The acute need to reexamine the current trends in biotechnology regulation and overregulation of GMOs is perhaps made most clear, ironically, by the perils of *under*regulating introductions of organisms that are not modified using rDNA techniques. Compared with the existing or planned regulatory schemes for genetically modified organisms, introductions of nonindigenous (that is, exotic) species often have been virtually unregulated, save for various national quarantine programs developed to deter parasites or other pests. Yet, the history of such introductions has been, at times, marked by disastrous consequences. Another reminder of the need for comparative risk assessment and management.

Accidental or deliberate addition of new species by human agency into foreign ecosystems has resulted in their serious disruption. It has often caused the extinction of native species, especially in island ecosystems: of those cases of historical extinction in which the cause can be identified, nearly 40% have resulted from the introduction of exotic organisms, mostly animals.[29]

In many cases the introductions are accidental, but often the introductions have been undertaken with a specific agricultural objective. Frequently it is the control of some undesired species, itself often of exotic origin. Some of the well-known examples form a chain of folly: one species is introduced, replaces native species and reproduces wildly, whereupon a second nonindigenous species is introduced in order to control it. The oft-told tale of rat and mongoose in the sugar cane fields of Hawaii is a familiar case, but by no means unique.

Unintended effects of biological control strategies have supplied the most dramatic illustrations of damage by nonindigenous species, but serious problems have also arisen from organisms imported for direct agricultural purposes. The Golden Apple Snail was brought to Asia in the early 1980s with the entrepreneurial promise that this fast-growing edible mollusc would be a valuable export commodity. It spread rapidly and soon infested rice fields in the Philippines, Korea, Thailand, Indonesia and Vietnam. There was little consumer interest in the product—Europeans did not like its taste—and its depredations on rice paddies have made a huge impact on the economics of that crop.[30]

There is obvious advantage in having a regulatory system that can address comprehensively the introduction of potentially hazardous organisms, whether "genetically modified" or not. But developing nations are being encouraged to consider as models for their own policies the technique-based biotechnology regulatory systems of the United States and Western Europe. These are not necessarily appropriate models for the social, scientific and regulatory milieus found in the developing countries, however, and they may not be compatible with existing resources or government infrastructure. In addition to the many concerns described above, the regulatory systems in industrialized countries often embody features that are overly risk-averse and ill-considered. They are often not even well suited to the nations that developed them. Consider, for example, an analysis of regulations in the United States which finds that the cost per life saved ranges from $100,000 to more than $5 *trillion*[31]—a range of more than seven orders of magnitude! It is difficult to imagine circumstances in which the latter, an EPA regulation, "Hazardous Waste Listing for Wood-Preserving Chemicals," could be judged socially or economically useful. The same is true of EPA, FDA and USDA regulations specific for rDNA-modified organisms.

The countries of the developing world should also be concerned about such regressive regulatory policies comprising their access to technologies, products and markets. A disproportionate regulatory burden on field trials of new biotechnology products is likely to have three relevant consequences: reduction of the overall number of field trials; discouragement of much needed public and private sector sponsorship of research; and, particularly in the longer term, limitation of the range of applications and potential domestic and export markets. All of these effects have already been seen in the U.S., Europe and Japan.

The dynamics among regulation, regulatory disincentives to R&D and reduced access to markets and products is determined by the way in which crop improvement occurs. The conventional genetic approach to crop enhancement has resulted in thousands of individual improvements in plants in order to overcome locally important constraints on crop production. These are typically "delivered" into practice in the form of locally adapted crop varieties—an approach that will not be available to new biotechnology if the newer molecular methods of manipulation are subjected to an assortment of "case-by-case-every-case" regulatory disincentives.

If they are to have access to the products of agricultural biotechnology, developing countries must make and implement their own reasoned decisions regarding appropriate regulatory structures. The approaches chosen by industrialized nations provide little useful guidance except, perhaps, viewed from a longer historical perspective—in other words, the usefulness of a vertical, rather than horizontal, approach to oversight (chapter 1, essay 5). The industrialized nations have been active in the development of organisms altered by pre-molecular techniques, including plants, microorganisms, mammals, fish and arthropods. Risks associated with the introduction and use of these new organisms have been effectively managed by governmental, professional and voluntary systems of policies and standards. As emphasized in the U.S. NAS and NRC reports, these historical approaches are both appropriate and adequate for managing the risks of new rDNA-modified organisms (see chapter 1).

Through January 1993, developing countries had accounted for only about 8% of the world's total field trials with recombinant DNA-modified plants.[32] There is a scarcity of recombinant DNA-derived organisms appropriate for developing countries' markets owing largely to low levels of research investment by both the public and private sectors. There are also concerns about the absence of regulatory systems for biotechnology research and products, fueled by fears that industrialized nations will "dump" dangerous products in developing countries. Some governments and Nongovernmental Organizations (NGOs) have called for the urgent and immediate implementation of biosafety regulations to ensure the safe conduct of biotechnology research.[33] As a condition of research support, some institutional donors, such as the Rockefeller Foundation, require regulatory approvals prior to field testing or commercial use of products.

Given the kinds and magnitudes of risks that actually exist, what kinds of biosafety mechanisms are needed and who should provide them? The next section describes the efforts of United Nations agencies to develop a global regulatory system. In chapter 5, I describe an algorithm for field trials[34] that, unlike the UN agencies' approach, is scientific and risk-based, could ensure safety and also enable developing countries to create and sustain their own local biosafety systems.

ESSAY 2
BIOTECHNOLOGY AND THE UN: NEW CHALLENGES, NEW FAILURES

UN agencies seem intent on becoming the world's "bio-cops" through the creation of sweeping safety regulations for both developing and industrialized nations. Unfortunately, the would-be superregulators appear to regard scientific consensus on issues of risk as an aspect of inconsequential policy making. UN officials' impact on biotechnology R&D and the use of new products by developing countries is likely to be profound and negative.

Of particular concern are two regulatory instruments: the UN Industrial Development Organization's (UNIDO) 1992 "code of conduct"[2] for field trials and the 1992 Convention on Biological Diversity[1] (CBD, or informally, the "Biodiversity Treaty"). The CBD's goals are laudable, but it has become the UN's Trojan horse, surreptitiously delivering biotechnology-averse regulatory policies to the developing world. The UNIDO code of conduct shares all the disadvantages of the CBD but none of its potential benefits.

The UNIDO Code of Conduct

UNIDO's proposal was published on behalf of three other UN agencies: the United Nations Environment Program (UNEP), the World Health Organization (WHO) and the Food and Agriculture Organization (FAO).

The preamble states that the purpose of this (currently) voluntary code of conduct is to "provide help to governments in developing their own regulatory infrastructure and in establishing standards" for research and use of GMOs—and then it proceeds to dictate regulatory requirements in the most stringent and unscientific terms. The document asserts that "(t)he UN is an obvious system through which to coordinate a worldwide effort to ensure that all [research and commercial applications of GMOs] are preceded by an appropriate assessment of risks,"[3]—but it demonstrates no grasp of the crucial risk analysis issues. The document also defines as a principal goal to "encourage and assist the establishment of appropriate national regulatory frameworks, particularly where no adequate infrastructure presently exists"[4]—but provides no economic strategies by which such frameworks could be created and sustained in developing countries.

The code requires the establishment of new environmental bureaucracies and demands resources from impoverished developing countries if researchers in those countries wish to perform even *small-scale field trials* of crops of local agronomic value, such as cassava, potatoes, rice, wheat and ornamental flowers. "Every member country should designate a national authority or authorities to be responsible for handling enquiries and proposals, i.e., all contacts concerning the use and introductions of GMOs. More than one authority may be necessary... Case-by-case evaluation should be the rule..."[5]

Finally, we get to the payoff for the international regulators-in-waiting themselves. The code proposes that the UN "establish an international biosafety information network and advisory service." These entities would have a number of functions, among them information-gathering, advice and technical assistance on monitoring the environmental impacts of GMOs, and "on request provide advice to assist in working toward the setting up of a designated national authority(ies) in each countryo"[6]

Because UNIDO has produced the code of conduct, it should not come as a surprise that "UNIDO should take the lead..." in establishing these new services. The resources required would include "a scientific steering committee...[and] a small technical/administrative secretariat." And "as a starting point, the service should conduct an international survey to identify existing expertise in the various scientific disciplines required for the safety assessment of biotechnology use."[7]

There is nothing that redeems this ill-conceived essay into regulation. Having gotten the scope of risk assessment completely wrong, the drafters of the code of conduct awkwardly invoke the language—but not the intent—of the scientific consensus about biotech regulation: "Regulatory oversight and risk assessment should focus on the characteristics of the product rather than the molecular or cellular techniques used to produce it."[8] They reduce that consensus to mere rhetoric forging ahead with a contradictory, expensive and regressive regulatory system. Thereby, they erect steep barriers to R&D in developing countries aspiring to meet some of their economic development and national security goals through biotechnology.

It is difficult not to dwell on better ways that resources could be spent than on this UN regulatory humbug: research on schistosomiasis, malaria and AIDS; immunizing children against cholera, polio and hepatitis; and developing genetically improved cultivars of staple crops like rice, potatoes and cassava.

The Convention on Biological Diversity

A product of the 1992 UN Conference on Environment and Development (UNCED), held in Rio, the CBD addresses a broad spectrum of issues related to the protection of biological diversity. The intention—conservation of habitats in developing nations—is laudable. And so are specific goals: (1) identifying and monitoring components of biological diversity, such as specific ecosystems and communities (Article 7); (2) establishing a system of protected areas (Article 8); (3) adopting measures for *ex situ* conservation (that is, preserving seeds or sperm in repositories under appropriate conditions); (4) integrating genetic resource conservation considerations into national decision-making and adopting incentives for the conservation of biological resources (Articles 10 and 11), and; (5) developing assessment procedures for ensuring that impacts on biological diversity are considered in project design (Article 14).[9]

The treaty was conceived as an unprecedented opportunity for industrialized and developing countries to reconcile issues of conservation and access to biological resources. However, it is burdened with a still unresolved international biosafety protocol (i.e., regulations for biotechnology) that would exacerbate rather than lessen challenges facing biodiversity, as discussed above in the section on developing countries.

The devil in the details

It is difficult to determine from the vague and sometimes contradictory language of the treaty whether the CBD actually requires that a biosafety protocol be adopted by ratifying countries. For example, Article 8 calls for measures to "regulate, manage or control the risks associated with the use and release of living modified organisms resulting from biotechnology which are *likely to have adverse environmental impacts* that could affect the conservation and sustainable use of biological diversity..."(emphasis added).[8] Article 19 specifies that "the Parties shall *consider the need for* and modalities of a protocol setting out appropriate procedures, including, in particular, advance informed agreement, in the field of safe transfer, handling and use of any living modified organism resulting from biotechnology *that may have adverse effect on the conservation and sustainable use of biological diversity*"[11] (emphasis added).

At the heart of the disagreement over the protocol is the meaning of the CBD's phrase, "any living modified organism resulting from biotechnology." Among field trials of living organisms including microorganisms, insects, fish, farm animals and crop plants and ornamental flowers, what, exactly, would be subject to regulation? All organisms tested in field trials? Certain organisms or field trials judged to be high-risk? The use of specific genetic techniques? The CBD's scope is potentially broad: "Biotechnology means any technological application that uses biological systems, living organisms or derivatives thereof, to make or modify products or processes for specific use."[12]

The advantage of such a broad definition is that, in theory, it provides latitude for circumscribing categories of organisms (to be used in a field trial) that deserve a government safety review—that is, a risk based approach focused on the product. Case-by-case review could be limited to those categories judged to be of possible significant risk. Such a risk-based approach[13] would be more defensible scientifically than the process- or technique-based approaches of the EU directive,[14] the UNIDO code of conduct,[15] or U.S. biotechnology regulation by EPA and USDA.

The rational possibilities afforded by the wording of the treaty are being overlooked. There is every indication that the international bureaucrats plan to do the wrong thing. An official document that describes the April 1993 proceedings of the expert panel established to implement these aspects of the CBD clarified the scope of the proposed regulations; according to paragraphs 57 and 58,

> a majority of the Panel members believed the organisms covered by a possible protocol should be restricted to genetically modified organisms, along the lines of the EEC-Directive 90/220 on the Deliberate Release into the Environment of Genetically Modified Organisms which defines GMO's as organisms in which the genetic material has been altered in a way that does not occur naturally by mating and/or natural recombination. Annex 1 A of the Directive further specifies which techniques are covered and which are not.[16]

The document states explicitly that the scope of regulation "does not include organisms modified by traditional breeding methods," regardless of pathogenicity, likelihood of constituting an environmental nuisance or other potential risks. The panel unabashedly endorsed an unscientific, process-based definition of what requires regulation that is wholly irrelevant to risk. It is noteworthy that they did so in spite of a minority viewpoint cited in the panel's report that argued against the GMO-limited scope. The dissenting view held that such a scope would be irrational and counterproductive because it "would ignore organisms actually known to present a threat to biodiversity, while focusing on others for which only hypothetical analyses can be offered."

The scientific view did not carry the day. Representatives from the EU, Holland and the United Kingdom labored to convince developing countries that their regulatory and development needs could be met by the same heavy-handed regulatory approach that currently impedes biotechnology research and commercialization in Europe. In characteristically paternalistic, if not imperialist fashion, they further offered to stand ready to perform the required environmental assessments for their lesser-prepared counterparts in developing countries.

These tactics deserve comment. The national representatives attending the CBD panels have been primarily *environmental* regulators, who, unlike regulators of products such as drugs and pesticides, commonly have not yet obtained first-class tickets on the bureaucrats' gravy train. It is not surprising, therefore, that a broader perspective on the negative consequences of unnecessary regulations was not presented. It is likewise not surprising that the resulting recommendations would aggrandize the participants' own professional responsibilities, budgets and influence. The outcome would likely have been quite different, had the attendees been science/technology or trade ministers or heads of state.

Negative impacts on foreign and domestic biotechnology

The CBD's wrongheaded regulation in the making will exert negative impacts on R&D that are not limited to the developing world, nor even to international transfers and transactions. Consider that in paragraphs 74 and 75 of the April 1993 panel proceedings report, there is the ominous observation that

> a majority of the Panel members interpreted the language of Article 19.3 'safe transfer, handling and use' as if both international transfer of organisms *and domestic handling and use* of organisms were covered. The majority of the Panel members thought that *domestic regulation should be covered by a possible protocol* (emphasis added).[17]

After several meetings and the production of prodigious amounts of paper by discussant panels, the nations that have ratified the CBD appeared to reach agreement on this issue at a conference in Jakarta in November 1995. At first, it appeared that the regulations would encompass both international (that is, transnational) and domestic research and development. Then, according to a November 20, 1995 report in the BNA [Bureau of National Affairs] Environment Daily, one of the key decisions made by the 1000-plus delegates at the Conference of the Parties was "an agreement to draw up a protocol governing the *transboundary* movement of living modified organisms"[18] (emphasis added). Thus, signatory nations agreed to formulate an international protocol regulating the transfer of GMOs *between states.* The agreement appeared not to extend to regulation of purely domestic testing or uses of organisms. The journal *Nature* reported, "The agreement represents a compromise between the Group of 77 (G77) countries, which favored early enactment of a comprehensive protocol on biosafety, and the industrialized nations, which have been seeking the adoption of looser guidelines."[19]

It is by no means clear that domestic regulation is beyond the reach of the protocol, however. The Indonesian Assistant Minister for the Environment, Effendi Sumardja, head of the working group that drafted the mandate, insisted that developing countries' demands for a broader protocol were consistent with the language in the compromise text. "We feel that transboundary movement of living modified organisms...is not specifically addressed by existing international agreements," Sumardja explained. "That's why we're focusing on this issue. It doesn't mean that the other elements (such as the safe handling and use of LMOs [live modified organisms, apparently synonymous with GMOs] are being deleted. We feel that it's still possible to put the other elements in our protocol."[20] (There's something bizarrely theosophical about all this. One can imagine the G77 banding together to insist that an agreement specify the number of genes that can dance on the head of a pin.) Again, an environmental regulator clamors for policies that will expand his own influence and bureaucratic empire.

The most recent document to emerge from this gratuitous exercise is the product of a meeting held in Cairo, December 11-14, 1995 by the United Nations Environment Program (UNEP). It is a 42-page treatise, entitled, "Report of the Global Consultation of Government-Designated Experts on International Technical Guidelines for Safety in Biotechnology."

The group adopted the "final text" of the UNEP Guidelines for Safety in Biotechnology and requested the UN to distribute them widely. Despite appearances to the contrary, this is *not* the definitive biosafety protocol. That will be produced by yet *another panel*, an "open-ended Ad Hoc Working Group to develop in the field of the safe transfer, handling and use of living modified organisms, a protocol on biosafety." But since it "may take several years to conclude the protocol...a number of countries and intergovernmental, private sector and other organizations will need technical guidance of the kind contained in these Guidelines to fulfill their commitments under such an international agreement."[21]

Whilst trying to appropriate the concept of "familiarity" from earlier work such as the 1989 U.S. NRC report and various OECD publications, the UNEP group in Cairo got it wrong, asserting that there is

> generally less familiarity with the behavior of organisms whose genetic make-up is unlikely to develop naturally, such as organisms produced by modern genetic modification techniques, than with the behavior of organisms developed

> traditionally. This has been the reason why many countries have focused on such organisms and products containing them...
>
> The guidelines provide assistance for identifying organisms whose characteristics may differ from those of the parent organisms from which they are derived in ways that would suggest additional scrutiny might be appropriate. This may be because they produce substances which are not found in the species concerned...[22]

These are precisely the rationales that the scientific community has repudiated consistently during the past decade. The UNEP document's misconceptions derive from confusion among various concepts, including *novelty, familiarity* and *risk.* The introduction of a new gene—even one in a location unlikely to occur in nature—does not necessarily affect an organism's relative risk. For example, a microorganism or plant may be newly carrying a biochemical or visual marker (such as *lacZY* or firefly lucifierase, respectively) introduced by rDNA techniques. If the host organism is known to pose negligible risk and the introduced gene does not make the new construct "unfamiliar" *with respect to risk,* then the modified organism should not fall into the regulatory net. Therefore, when field tested, it should be treated no differently than organisms with similar traits modified by traditional techniques (which are generally unregulated, unless they are known or suspected plant or animal pathogens, etc.).

The folly of UN officials (one hesitates to say "scientists") seems boundless. At meeting after absurd meeting, the United Nations builds an unstable edifice: misstatements are heaped upon misapprehensions, all on a foundation of scientific illiteracy. (UN policy makers seem to rely heavily on the old Washington D.C. political standard that anything said three times becomes a fact.) Unfortunately, these antics have real costs for citizens of the UN's member nations. They pay for the direct expense of these recurring international seances, as well as the indirect expense of ultimate compliance with the new regulations.

The United States Senate must ratify the CBD if the United States is to be a party to it and subject to the biosafety protocol. This is unlikely to happen while the Republicans control the Senate, which has final responsibility for ratifying treaties.

Some observers have interpreted the presumed requirement for Senate consent to a biosafety protocol as a fail-safe against U.S. participation in an imprudent agreement. However, were the CBD to be ratified by the Senate, the Clinton administration would not be required to seek its advice or consent to any of the new language developed by a conference of the ratifying nations and agreed to by the administration. Moreover, even if the Senate were consulted, various blandishments and pressures could be brought to bear on them by the administration and others. This latter scenario is of particular concern with the CBD high on the administration's agenda. Speaking at Stanford University on April 26, 1995, Undersecretary of State for Global Affairs Tim Wirth said that the CBD has the "top priority among all treaties" and agreements awaiting confirmation. It is also a pet project of Vice President Gore, who has shown both an intolerance of biotechnology and an environmental fervor characterized more by bombast than biology (see chapter 2).

An additional issue in the context of the biosafety protocol negotiations described here is which principles or positions are appropriate for compromise? "Compromises" pushed by bureaucrats (and sometimes by industry, see chapter 5) remind me of a cartoon that is especially apt in this international context. It depicts a primitive prehistoric warlord conversing with his lawyer. The latter has just informed his client that he's accused of pillaging and plundering Paris. "What'll we do," asks the worried client. "I'll try to get it reduced," the lawyer promises. "To what?" "Pillaging and plundering Helsinki," says the lawyer.

Far too often, portentous issues of public policy endure such compromises. Bureaucrats bargain. They use appropriate scientific buzzwords, seductively, while concealing a flawed scientific paradigm. They encourage all parties to agree to moderate against the most "extreme" positions. In the end, however, they settle on highest common denominator regulatory burdens. On scientific issues, the result of "splitting the difference" is usually unsatisfactory and often bizarre.

Conclusions

It is ironic that both sets of UN regulations (i.e., the UNIDO code of conduct and CBD) would stifle the development of environment-friendly biotechnology innovations that can help increase food production, clean up toxic wastes, purify water and replace agricultural chemicals. With such poorly crafted regulatory proposals, the developing world has been relegated to a lose-lose position. New-biotechnology processes and products are made artificially expensive to test, produce and use. Unnecessary case-by-case review diminishes the diffusion of useful new technology to the developing world. Developing countries are prevented from participating in worldwide technological trends. The technology gap between industrialized and developing countries widens. The income gap between the "have" and "have-not" nations, which more than doubled between 1960 and 1991, widens further.

There are bitter ironies in the demands that developing countries adopt unscientific, technique-based and anti-innovative regulatory approaches. Regulators and environmentalists from industrialized countries regard third-world regulation as an opportunity both to validate their minority view of biotechnology risks and to advance their own importance as policy players on a larger stage. The policies they favor will systematically undermine research on precisely the kinds of products that are most needed in developing countries—more plentiful and nutritious foods and biological alternatives to chemical pesticides and fertilizers. At the same time that their poorly conceived policies afford no increment in environmental or public health protection, they exact an unacceptable opportunity cost in the form of deferred enhancements in food production, health care and entry into international markets.

Some observers, such as Stanford Law Professor John Barton, have argued that the United States can best oppose the biosafety protocol by ratifying the treaty and then tenaciously objecting to the protocol while "participating as a full player at the table."[23] However, my experience in international negotiations and with the climate of these discussions suggest that U.S. negotiators will be hamstrung—always on the defensive, beleaguered by an irrational, paranoid and angry coalition. Consider a senior USDA participant's impressions of the June 1994 deliberations on the biosafety protocol by the Intergovernmental Committee for the Convention on Biological Diversity:

> "widespread ignorance about biotechnology among developing country delegates, coupled with fears based on past experience with [dumping of unsafe products] from the North;" a "contentious and polarized climate," in which

> the U.S. views were isolated and demonized; "grotesque" and "revisionist" misrepresentations by a certain developing country about previous consensus on the need for a biosafety protocol; and "rabidly antibiotechnology propaganda by three Nongovernmental Organizations (NGOs)... [which] introduced a series of antibiotechnology canards, misrepresentations and distortions as factual," assertions that were "taken as gospel by the legions of uninformed."[24]

By ratifying the CBD, the United States would commit itself to negotiating in an intellectual and political miasma. Article 23 of the CBD, in effect, limits eligible NGO participants to only the "greenest" of the environmentalist groups. NGOs can be excluded if one-third of the countries present object to the admission. This precludes the participation of organizations that represent commercial mining, timber, agribusiness, livestock, fishing and energy interests. It is ominous that in Jakarta the more radical NGOs such as Greenpeace and Friends of the Earth—and only such groups—were permitted to participate in the usually closed negotiations. It should come as no surprise that the Jakarta result was neither rational nor scientifically defensible.

The biosafety debacle detracts from the CBD's otherwise laudatory biodiversity goals. Stanford University biologist Donald Kennedy has emphasized eloquently that each vanishing species is an irreplaceable resource, the unique end-point of a long evolutionary process that cannot be duplicated. "One may be the keystone of a complex ecosystem that plays some role in climate stabilization. Another contains gene sequences that may yield new understanding about the control of behavior. Still another may, because of its beauty or its grandeur, have a special hold on the human imagination—as eagles have for some of us, or otters for others."[25]

One of the CBD's problems lies in the mechanisms by which its worthwhile goals would be implemented. They are variously ambiguous, vague or impotent. The greater problem is the biosafety protocol. Its ability to enhance ecological safety in the developing world is extremely doubtful. The likelihood of it being cost-effective is nil.

If the UN insists on having a role in international regulation, its mandate and its policies must make economic and scientific sense. Thus far, the UN has failed to meet any of these conditions.

References

Essay 1: Real World Effects

1. Breyer S. Breaking the Vicious Circle. Cambridge: Harvard, 1993:23.
2. Wildavsky A. Wealthier is healthier. Regulation 1980; 4:10.
3. Keeney RL. Mortality risks induced by economic expenditures. Risk Analysis 1990; 147:148-49. See also Catalano R. The health effects of economic insecurity. Am J Public Health 1991; 81:1148.
4. Taylor J. Salting the Earth—The Case for Repealing Superfund. Regulation 1995; 2:53-66.
5. Miller HI, Concepts of Risk Assessment: The "Process Versus Product" Controversy Put to Rest. In: Brauer D, ed. Biotechnology. Weinheim: VCH, 1995:39-62.
6. Anon. Biotechnology Risk Control Luxembourg: Office for Official Publication of the European Communities, 1994.
7. Anon. European BioNews October 1992; 1:1.
8. Anon. Impact of Genetic Engineering Regulation on the West German Biotechnology Industry. RauCon Biotechnology Consultants GmbH, Dielheim, Federal Republic of Germany, 1989.
9. Abbott A. Transgenic trials under pressure in Germany. Nature 1996: 380:94.
10. Anon. Biotechnology Risk Control Luxembourg. Ibid.
11. Ward M. Do U.K. regulation of GMOs hamper industry? Bio/Technology 1993; 11:1213. See also, Young FE, Miller HI. Deliberate releases in Europe: over-regulation may be the biggest threat of all. Gene 1989; 75:1-2.
12. Anonymous. Biotechnology in a Global Economy. U.S. Congress, Office of Technology Assessment. OTA-BA-494. Washington: U.S. Government Printing Office, October 1991.
13. Anon. Report on Regulation of the United Kingdom Biotechnology Industry and Global Competitiveness. HL Paper 80. Her Majesty's Stationery Office, 1993.
14. Anon. European Union Modifies BST Moratorium. Bioworld Today Dec. 19, 1994; 2.
15. Anon. Regulatory Body Rejects Approval of Genetically Modified Maize Seed. International Environmental Reporter May 1, 1996.
16. Idem.
17. Craig C. European Union Officials Worry U.S. Biotech Lead Becoming Insurmountable. BioWorld January 24, 1996; 1.
18. Anon. Guidelines for Foods and Food Additives Produced by Recombinant DNA Techniques, Ministry of Health and Welfare, Government of Japan, 1992.
19. McNicoll G. The United Nations' long-range population projections. Pop Devel Rev 1992; 18:333-340.
20. Idem.

21. Udelman M et al. Feeding 10 Billion People in 2050: A Report by the Action Group on Food Security. Washington D.C.: World Resources Institute, 1994.
22. Anonymous. FAO Yearbook. Rome: United Nations Food and Agriculture Organization, 1992.
23. Chrispeels MH, Sadava DE. Plants, Genes and Agriculture. Boston: Jones and Bartlett, 1994, chapter 15.
24. Dyson T. Population Growth and Food Production: Recent Global and Regional Trends. Pop Devel Rev 1994; 20:397-411.
25. Smil V. How Many People Can the Earth Feed? Pop Devel Rev 1994; 20:255-292.
26. For a discussion of these prospective limitations, see Ehrlich PR, Ehrlich, AH, Daily GC. Pop Devel Rev 1994; 19:1-32.
27. Dasgupta P. Population, Poverty and the Local Environment. Scientific American 1995; 41-45.
28. Waggoner PE. How Much Land Can Ten Billion People Spare for Nature? Ames: Council for Agricultural Science and Technology, 1994.
29. Anon. Global Biodiversity: Status of the Earth's Living Resources. Report of the World Conservation Monitoring Center. London: Chapman and Hall, 1992:199.
30. Naylor R. Invasion of Exotic Species in Agriculture: A Cautionary Tale of the Golden Apple Snail in Asian Rice Systems, Ambio; (in press).
31. Breyer S. Ibid.
32. Goy PA, Chasseray E, Duesing J. Agro Food Industry Hi-Tech 1994; 5:10.
33. Kaiser J, Science 1994; 266:1935. Also Hughesman M. BioWorld Today January 9, 1995; 2.
34. Miller HI, Altman DW, Barton JH et al. Biotechnology Oversight in Developing Countries: A Risk-Based Algorithm. Bio/Technology 1995; 13:955. Also Miller HI, Burris RH, Vidaver AK et al. Risk-based oversight of experiments in the environment. Science 1990; 250:490.

Essay 2: UN

1. Anon. Biotech Forum Eur 1992; 9:218-21.
2. Anon. Convention on Biological Diversity, United Nations Environment Program (UNEP), 1992.
3. Anon. Biotech Forum Eur. Ibid.
4. Idem.
5. Idem.
6. Idem.
7. Idem.
8. Idem.
9. Anon. Convention on Biological Diversity. Ibid.

10. Idem.
11. Idem., emphasis added.
12. Idem.
13. Miller HI, Burris RH, Vidaver AK et al. Risk-based oversight of experiments in the environment. Science 1990; 250:490. See also Miller HI, Altman DW, Barton JH et al. Biotechnology Oversight in Developing Countries: A Risk-Based Algorithm. Bio/Technology 1995; 13:955.
14. Anon. Biotechnology Risk Control. Office for Official Publications of the European Communities, Luxembourg, 1994.
15. Anon. Biotech Forum Eur. Ibid
16. Anon. Expert Panels Established to Follow-Up on the Convention on Biological Diversity, Report of Panel IV, UNEP/Bio Div/Panels/Inf. 1, April 28, 1993.
17. Idem., emphasis added.
18. Anon. BNA [Bureau of National Affairs] Environment Daily on November 20, 1995.
19. Masood, E. Biosafety rules will regulate international GMO transfers. Nature 1995; 378:326.
20. Idem.
21. Anon. UNEP/Global Consultation/Biosafety/4. December 19, 1995.
22. Idem.
23. Barton JH. Biodiversity at Rio. BioScience 1992; 42:773.
24. Giddings LV. USDA Trip report, July 1994.
25. Kennedy D. Don't Be Distracted by Alien Cows. The Los Angeles Times, October 24, 1994; B9.

CHAPTER 5

WHY GOVERNMENT IS THE PROBLEM (AND WHAT TO DO ABOUT IT)

INTRODUCTION

P.J. O'Rourke, the political satirist and author, has said that giving money and power to government is like giving whiskey and car keys to teenage boys. I agree, but he raises a central question addressed from many perspectives in this volume: *Why* is government the problem?

There are a number of reasons. At the root of many is bureaucratic and commercial self-interest in conflict with the public interest.

In this chapter I consider both the relationship between the biotechnology industry (as represented by its trade associations) and federal agencies; and the system of bureaucratic incentives, disincentives, rewards and punishments within the FDA and EPA that affects officials' decision-making.

There are some possible solutions but there is no magic bullet or instant formula. Some extend from first principles, others from empirical experience. Some are sweeping and attempt to introduce systematic changes, others are specific and mechanistic. The point is—change is both needed and possible. This chapter presents constructive proposals.

ESSAY 1
SELF-INTEREST AND THE FORMULATION OF PUBLIC POLICY

ONEROUS REGULATORY POLICIES? BLAME INDUSTRY!

America learned long ago that what's good for General Motors isn't necessarily good for the country. This axiom applies equally well to the biotechnology industry. For years a triad of industry titans (the Monsanto Company, Ciba-Geigy Corporation and Pioneer Hi-Bred International) have dominated the biotechnology industry's trade associations. It hardly sounds plausible that industry would wish to crank up the level of regulation, but just consider how they stand to benefit from regulation that acts as a market-entry barrier. Such barriers can be a potent weapon against smaller, highly innovative competitors.

The biotechnology industry has not driven a single major regulatory improvement in the past 20 years. They have been, at best, tentative on a variety of regulatory issues. Since the mid-1980s, biotechnology trade associations have lobbied feebly, accepted—and even proposed—policies and compromises that disadvantage their own members in the competitive marketplace. They have often shown a myopic fixation on creating governmental assurances for environmental and consumer activists, at any cost to innovation and competition. But this myopia is neither the only nor the best explanation. In a Faustian bargain, overregulation gives big companies market-share protection and shelter for existing products while government bureaucrats get bigger budgets and empires. Highly innovative small and mid-sized companies and the nation's academic research enterprise find R&D impeded by excessive costs and delays. American innovation and good old-fashioned competition go to hell.

Biotechnology trade groups' checkered past

A few examples will illustrate how the biotechnology trade associations served the interests of a few big and influential companies in an anticompetitive campaign inimical to the public interest.

In the mid-1980s, North Carolina implemented a new law that created a new state entity specifically to regulate field trials of rDNA-modified organisms. The new regulations required extensive evaluation of testing of common crops such as corn, wheat, tomatoes and tobacco. The state's rules were both scientifically flawed and superfluous to public and environmental health. Even the state regulators involved with their creation and implementation were apologetic about them.

Representing the industry at the time were the Association of Biotechnology Companies (ABC) and the Industrial Biotechnology Association (IBA), which later merged to form the single Biotechnology Industry Organization (BIO). To the bewilderment of the biotechnology community, the ABC endorsed the North Carolina law. An ABC official observed privately that his organization, vying with the IBA for visibility and members, had to do something—anything—to distinguish itself. It succeeded.

Another episode occurred during a 1986 public meeting of the federal interagency Biotechnology Science Coordinating Committee. The BSCC had been established to, among other goals, enable various biotechnology stakeholders and interest groups to have a voice in policy decisions. At the time, the federal agencies were struggling with the question of the "scope of regulation," that is, whether regulation should focus on the use of gene-splicing techniques or on the risk-related characteristics of products. At the 1986 meeting, ABC and the Pharmaceutical Manufacturers Association (PMA) described the enormous investments that their members routinely made in R&D and the importance of their new products to consumers and the U.S. economy. They argued for a scientifically sound, risk-based and predictable approach to regulation. Poorly conceived and burdensome regulation, they said, could put whole industrial sectors and tens of thousands of jobs at risk. Then, the IBA's representative took the podium. He told the assembled audience, including TV cameras, that his members were tired of discussing regulation: what companies really needed from government was—more databases. *Databases?* One person in the audience was so astonished that he literally fell out of his chair.

These bewildering incidents were typical of the trade associations' contributions to the debate in the 1980s. The associations were characteristically unreliable whenever Reagan and Bush administration officials requested their support for scientifically defensible regulation at EPA and USDA.

The BIO monolith

BIO, which now represents the ostensibly coalesced national interests of the U.S. biotechnology industry, has pursued a strategy defined largely by myopia and low expectations on major issues. Its communications with the Clinton administration have typically focused on resource issues. With two exceptions, regulation has fallen from BIO's agenda. The two exceptions are FDA reform and the Convention on Biological Diversity (CBD, or more colloquially, the Biodiversity Treaty).

Consider the recent BIO proposals for FDA reform. It was only after relentless attacks on the FDA in major newspapers,[1] the announcement of several reform initiatives by think-tanks,[2] and the publication of a detailed reform proposal[3] that BIO finally was compelled to address the need for reform. By contrast, European companies have consistently worked through the Brussels-based Senior Advisory Group, Biotechnology, exerting pressure to revise EU regulations. The slow-moving BIO finally released its FDA reform proposals in early 1995 but offered little beyond what the Clinton administration and the FDA had already conceded (see chapter 3).

BIO's most serious policy failings have been in the realm of EPA and FDA regulation, and the CBD. R&D on bioremediation and pest-resistant garden and crop plants are the early casualties. A report on the evolution of U.S. biotechnology prepared for Canada's National Biotechnology Advisory Committee even listed Carl Feldbaum, BIO's president, among those promoting the "guilty until proven innocent" view of biotechnology product regulation.[4] BIO's position does not bode well for the entrepreneurial agricultural and environmental biotechnology companies. The association has actually been hostile to innovation and small business development in these emerging sectors (*vide infra*).

BIO and the EPA

Consider the cozy alliance between the powerful BIO and the EPA. For a decade, new regulations under the EPA's pesticide and toxic substances statutes have sidestepped scientific advice in order to single out for special scrutiny new biotechnology products that could supplant chemical pesticides and clean up toxic wastes. Scientific and professional associations such as the American Society for Biochemistry and Molecular Biology and the American Society for Microbiology, and the University of California's Systemwide Biotechnology Program, have criticized the EPA's general approach and its specific proposals as being unscientific, barriers to early stage research and, ultimately, contrary to the public interest. A common criticism has been that agency policies impede the progression from publicly funded research to tangible public benefits. (See, also, the discussion in chapter 3 of a report by 11 scientific societies which criticizes EPA's plant-pesticide policy.)

BIO, in contrast, has not joined the denunciation. In fact, it has consistently endorsed and defended the EPA's proposals including those (under both the pesticide and toxic substances statutes) that single out small-scale field trials of rDNA-manipulated microorganisms for regulation while exempting all others. In return, EPA policies have been crafted in a manner that allows exemptions from regulation for certain specific (and seemingly arbitrarily-selected) products—plants modified by the addition of a viral coat protein gene, for example. Such actions by EPA selectively benefit those companies far enough along in R&D to have gotten their products anointed and which possess sufficient regulatory experience to recognize the opportunity. Other companies, most of which are small and in earlier stages of R&D, are left behind to struggle with the new regulatory barriers.

BIO even supports the EPA's bizarre 1994 proposal to expand regulation of chemical pesticides to include *whole plants* manipulated with rDNA techniques—products which would enable farmers to use smaller amounts of chemical pesticides while sustaining crop yields. (The final rule still has not been issued; presumably, EPA intends to delay until after the November 1996 elections, which could change the now Republican-controlled Congress.) There is no question of this proposed rule serving any useful function. New pest-resistant plant varieties selected or crafted by older and less precise techniques have, after all, a long history of safe use—without any governmental regulation, let alone regulation as *pesticides*.

In 1995, BIO lobbied Congress feverishly to defeat a pivotal legislative rider, the so-called Walsh amendment, which would have denied EPA funding where "whole agricultural plants [are] subject to regulation by another federal agency" (see chapter 3). The Walsh amendment would have both limited the agency's expanding influence and shifted the regulation of many of these products under the FDA's risk-based 1992 policy on new-biotechnology foods.[5] Unblushingly, BIO asserted that they and their companies had "enjoyed a good working relationship with EPA" and described EPA officials as "flexible in their regulatory approach" and as being eager to minimize the impact of regulation on biotechnology.[6]

The amendment passed the House by a 210-210 procedural vote despite the attack. At the time, a spokesman for BIO promised that the association would continue the battle in the Senate. The Senate subcommittee chairman responsible for the EPA appropriation was Christopher S. Bond, a Republican from Missouri, Monsanto's home state. The Senate subcommittee did not take up the amendment, and, to no one's surprise, Bond saw to it that it died in the House-Senate conference.

In their lobbying against the Walsh amendment, BIO and Monsanto worked closely to court key congressmen and strong-arm supporters of the amendment. An official of the Institute of Food Technologists, a nationwide professional association that supported the amendment, described their experience with BIO and Monsanto:

> I just had a meeting this morning with a delegation from BIO and Monsanto who trooped into town to educate me. All went well for the first hour and then, as time was nearing a close, they got down to the brass tacks of our [pro-amendment] position on the Walsh amendment. Still in my 'sweet' finding-common-ground way, I indicated that many [in our organization] supported our position and that we did not see any way to yield on that position. Bob Harness, VP for Registrations and Regulatory Services [at Ceregen, a Monsanto subsidiary] converted to Mr. Hyde when I indicated that we saw no reason to adjust our views. At that, he rose from the table, beet red, and declared that they would continue to fight us.[7]

Before leaving, the visitors demanded lists of the Institute's financial supporters.

BIO has thus been lobbying for new and duplicative regulation of plants by the EPA. It is significant that the BIO-favored EPA approach creates new barriers to commercial success for smaller, more cash-strapped companies and for academic researchers. The proposed plant-pesticide policy would discourage fundamental university research on the nature of pest-host interactions needed to develop biological pest control products. The most pernicious effect of BIO's approach is that it treats academic research as expendable. Some in industry seem to have forgotten that the genesis of the new biotechnology was the confluence and synergy of basic research in academic biochemistry, microbiology and analytical chemistry. Alternatively, do BIO members consider publicly funded research to be a competitive threat? Either way, BIO's cynical strategy undermines the foundation of U.S. competitiveness in high technology—viz, federally supported R&D.

Devotees of conspiracy theories may find it interesting that the implausible EPA plant-pesticide policy was crafted by Bush administration EPA Assistant Administrator Linda Fisher, who passed through the government's revolving door and is now vice president for government relations at Monsanto. It is noteworthy, too, that the large agricultural companies that control BIO are among the world's largest producers of chemical pesticides and control significant market shares of the pest control market. (Monsanto's sales of crop protection and lawn-and-garden products in 1995 were $2.47 billion, for example.)

BIO and the FDA

BIO and the Clinton regulators have often danced cheek-to-cheek. In 1993, BIO was an ally in FDA's plan to require the premarket registration of new biotechnology foods (see chapter 3). This single change would disrupt a successful and progressive 15-year FDA policy of treating biotechnology products the same as similar nonbiotechnology products. The principal beneficiaries would be only those who wish to slow the development of the new biotechnology products. The July 18, 1994 *BIO Bulletin* claimed that the association was playing hardball with the FDA, by accepting premarket registration while rejecting premarket approval. But premarket approval had never been a serious option.

Experts in plant science, food and nutrition persistently criticized the proposal for a premarket registration requirement. Following the November 1994 elections, the new Republican Congress made pointed inquiries about the FDA's intentions. Then, suddenly, BIO had a change of heart. On January 5, 1995, BIO withdrew support from the proposal. The Clinton administration followed suit, and shortly thereafter the FDA quietly deleted the food biotechnology registration policy from its regulatory agenda. Calling it "completed," the agency buried the announcement in a single line in an OMB publication.[8] It will be instructive to see BIO's response to the newly-resurrected FDA policy of discriminating against biotech-derived foods (chapter 3, essay 1).

BIO and the convention on biological diversity

While the primary goals of the CBD are laudable (but of little conceivable advantage to U.S. industry), the implementation is vague; worse still, the panels that have met to craft and implement the biosafety protocol mandated by the treaty have imployed regressive, unscientific approaches, singling out products made with rDNA techniques. BIO's vacillation on the CBD (chapter 4) is another sorry episode. After initially aligning themselves with the Bush administration's opposition to the CBD, BIO flip-flopped (with the change in presidential administration) and became a major booster. According to environmental writer Russ Hoyle, BIO may now have reversed course again:

> Thanks in part to the efforts of U.S. observers [of the drafting of the biosafety protocol] who have witnessed the campaigns of disinformation and bald deceit employed by the forces arrayed against U.S. biotechnology interests, [BIO] has at last begun to bestir itself for the debate over the biosafety protocol.[9]

BIO deserves few kudos for finally coming around on this issue. Reading the leaked copies of the draft biosafety protocol (prepared by a Washington D.C. "public interest" group called the Community Nutrition Institute), BIO must have begun to feel like Dr. Frankenstein with his monster running amok. Again, according to Hoyle,

> Its preamble declares that genetic diversity "is dependent on the socioeconomic conditions of the peoples maintaining it," code for a regime that includes social and cultural studies, sociology and "history relevant to risk assessment" in its definition of science. It designates illegal traffic in genetically engineered goods as a criminal act and includes jail sentences for responsible corporate and national

> officials. Besides requiring exporters of biotechnology products to submit complete safety information on a case-by-case basis, the draft protocol establishes an independent international body of experts to conduct risk assessments and make decisions on all transboundary trade in [rDNA-manipulated organisms].[10]

Follow the self-interest

Congressional committees with jurisdiction over these agencies and issues have been nonplussed by the goings-on. But, as Milton Friedman counsels, look for the self-interest. Sometimes the self-interest of industry, and often that of government regulators, is inimical to what is best for the rest of us.

Who benefits from the EPA's policies? The EPA grows in size and power, as it spawns new bureaucracies for regulating rDNA-derived organisms; and the big agricultural chemical/biotechnology companies thrive. BIO continues to be the darling of its largest, most influential members and reinforces the illusion of influence through its obsequious relationship with the Clinton administration.

Who loses? Small corporate competitors in biotechnology struggle against artificially high market entry barriers; several companies, in fact, (including DeKalb, Agracetus and Calgene) have recently been purchased, in whole or part, by Monsanto, Pioneer and Ciba-Geigy. The research community is discouraged by regulatory hurdles and shrinking resources; and ultimately, farmers, consumers and those interested in toxic waste cleanup are left with continuing problems but no new solutions.

The complex nature of BIO's self-interest, even beyond the immediate interests of its dominant members, is apparent in its actions. First, the strategy of lowered expectations and demands enables the organization to claim almost any outcome as a BIO victory. Second, BIO has cultivated an aura of "influence," even as it bows to government politicos, regulators and even the antibiotechnology troglodytes, while consistently eschewing efforts to do the right thing. Finally, the BIO leadership has listed noticeably to the political left. The following example is illustrative of BIO's shortcomings.

In 1995, BIO decided on a major new policy initiative. What was it? Something like transferable tax credits for research spending, so that its constituents, even small start-up companies without revenues, could receive tax benefits? Deregulation? Nope. *Bioethics!* In an August 4, 1995, letter to BIO's members, President Carl Feldbaum explained:

> As our industry progresses, bioethics issues grow increasingly important and they are frequently raised by the media in response to new biotechnology discoveries and developments. Currently, we are facing gene patenting issues, but the debate looms much broader and deeper. Ethical questions are also raised about gene therapy, transgenics and privacy and discrimination in genetic testing. Accordingly, BIO has designated bioethics as a *long-term top priority* (emphasis added).

Choosing bioethics as a top policy priority is like throwing a computer terminal to a drowning man. But it is characteristic of BIO: bioethics is a subject likely to be of low priority for most small and medium-sized companies, who worry most about product approvals, financing and cash flow; it elicits minimal expectations from member companies; it creates a great deal of "busy work" for BIO staff and consultants; and it enables BIO to claim success from even the most meager, insubstantial result.

Individual biotechnology companies and trade associations can and should take an aggressive position on federal regulatory policies that will shape the future of the technology and the industries that use it. The long view of regulation should be dictated by scientific principles, economics and common sense. It should resist the temptation of flawed, short-term fixes.

My solution to BIO's skewed priorities? Link the financial remuneration of BIO's top officials to tangible improvements in federal policy. These would include regulatory improvements at the government oversight agencies, as well as tax credits for research, patent reform, and so forth. The association's officials should get no credit for trying hard, staying late at the office, getting President Clinton to speak at the annual meeting, flowery oratory, slick press releases, voluminous reports or scintillating panel discussions. After all, BIO's mandate is to represent members' collective interests in the nation's capital. BIO officials' bottom line should depend on the legislative and regulatory bottom line. (I'd bet the mortgage money that this innovation would change BIO's priorities and level of effectiveness overnight.)

BUREAUCRATIC VERSUS SOCIETAL RISK

Government regulators, like the rest of society, generally act in their own self-interest, even when those actions are inimical to the best interests of others. The FDA and EPA both offer examples of self-interest clashing with the public interest. It's no secret that many EPA and FDA officials have a zero-risk mind set and believe they have a blank check to protect public safety. Aren't these same people eager to approve safe and innovative new products? Don't they want to encourage the development of improved ways to control pests, clean up toxic wastes and treat diseases? As a matter of fact, no.

They have other things on their minds. Civil servants and political appointees think a lot about simply staying out of trouble.

There is a marked asymmetry between the two kinds of mistaken decisions that a regulator can make that would get him into trouble: (1) a harmful product is approved for marketing (a "Type 1" error) or (2) a useful product that treats disease or promotes environmental protection is rejected, delayed, or never achieves marketing approval (a "Type 2" error).[11] In other words, the regulator commits a Type 1 error by permitting something harmful to happen and a Type 2 error by not permitting something beneficial. The consequences for the regulator are very different in the two cases.

Consider two examples of a Type 1 error—the FDA's approval of the "swine flu" vaccine, which caused temporary paralysis in a significant number of patients, and the EPA's approval of a pesticide that causes birth defects in humans and endangered species of birds. These kinds of mistakes are made highly visible by the media and are denounced by the public and legislative oversight bodies. Both the developers of the product and the regulators who allowed it to be marketed are excoriated and punished: modern-day pillories include congressional oversight hearings, CBS' *60 Minutes* and editorials in the *New York Times*. A regulatory official's career might be damaged irreparably by his good-faith but mistaken approval of a high-profile product.

Type 2 errors, in the form of unreasonable governmental demands or requirements, can delay or prevent entirely the marketing of a new product. But a Type 2 error caused by a regulator's bad judgments, timidity or anxiety usually gains little or no attention outside the company that makes the product. And if the regulator's mistake precipitates a company's decision to abandon the product, that is seldom widely known. There may be no direct evidence that continued patient suffering, farmers' loss of crops to insects or the reliance on outmoded technology for cleaning up oil spills is avoidable (chapter 3)—or that regulatory officials are culpable. The only counter-example is where activists closely scrutinize agency review of certain products and aggressively publicize Type 2 errors—such as the AIDS activist groups that monitor FDA.

Agencies and reviewers frequently justify and accept Type 2 errors, remonstrating that they are merely "erring on the side of caution." Too often this euphemism is accepted uncritically by the media and the public.

Consider, however, the possible societal impact of Type 2 errors. As related in chapter 3, the Monsanto Company several years ago proposed a scientifically interesting and potentially important small-scale field trial—biological control of a voracious corn-eating insect. The experiment would have used a harmless soil bacterium, *Pseudomonas fluorescens,* into which, using new biotechnology techniques, scientists had introduced a single gene from another, equally innocuous bacterium.

In spite of the unanimous conclusion of the EPA's panel of extramural scientific experts and other federal agencies that there was virtually no likelihood of significant risk in the field trial (and leaving aside the enormous potential benefit to farmers and consumers), the EPA refused to permit it.

Two aspects of this prototypic Type 2 error are so striking that they bear repeating: the field trial would have been subject to *no government regulation at all* had the researchers used an organism with identical characteristics but crafted with less precise "conventional" genetic techniques instead of the new biotechnology, and Monsanto's response to the rejection was to dismantle its entire research program on microorganisms for pest control.

In the ensuing decade, few other companies have pursued these products and dared to test the regulatory waters. This whole sector of biotechnology has been damped by regulatory disincentives—ironically, at a time when new markets have appeared as competing products (chemical pesticides) have fallen into societal and governmental disfavor.

Egregious and costly Type 2 errors can also take the form of broad agency policies. A good example is the 1994 EPA proposal to regulate as pesticides whole plants whose resistance to pests, disease or environmental stress has been enhanced by new biotechnology (discussed in chapter 3). EPA intends to regulate these plants, such as corn, tomatoes and marigolds, *more stringently* than chemical pesticides similar to DDT or parathion. You don't have to be a scientist to know that this makes no sense, affords no added protection to human health or the environment, and discourages research.

Many Type 2 errors are not so readily apparent, however. Rather, they reflect more the "culture" of risk-aversion in which every decision, every choice is overly conservative. For example, in the early 1980s when I was at FDA, the agency confronted an interesting decision about a new vaccine.

The first-generation vaccine for hepatitis B, which had been on the market for several years, was not a popular product. It was purified from the pooled plasma of patients with chronic active hepatitis, a population likely to be harboring many dangerous pathogens. Even though each batch of the vaccine was exhaustively inactivated, tested and purified, it was not used enthusiastically by physicians or patients.

The manufacturer of that product, Merck and Company, also developed the second generation, rDNA-derived vaccine. It had its origins in baker's yeast, *Saccharomyces cerevisiae*, which was modified by the addition of a single surface antigen gene from the hepatitis B virus. During fermentation, the yeast produced large amounts of the viral antigen which was highly purified. In this case the likelihood of contamination of the vaccine with human pathogens is virtually nil.

Demonstration of safety for this vaccine was straightforward. FDA's more vexing decision concerned efficacy—specifically whether the manufacturer had to perform clinical trials to show that the product actually prevented hepatitis, or whether a laboratory surrogate would be adequate. Arguably, it would be adequate to demonstrate that vaccine recipients synthesize the appropriate amounts and types of antiviral antibodies (a great deal was known about this "seroconversion" from the first-generation vaccine).

FDA's decision was important. Large amounts of time and money were at stake. Clinical trials to demonstrate hepatitis prevention had to be done in high-risk populations which are only found abroad, primarily in Asia; and organizing, performing and analyzing the studies would take years and be very costly.

In the end, FDA opted for the full clinical trials, rejecting even a middle course where seroconversion would be the primary measure of efficacy but a pilot study in Asia would confirm hepatitis prevention. The result was several years' delay while in the United States tens of thousands of cases of hepatitis B occurred annually that could have been prevented by the vaccine. (Approximately 5% of hepatitis B cases have serious complications and 0.1% are fatal.)

In my experience, in the current climate of regulatory agencies' risk-aversion and fear of Type 1 errors, judgments like this one occur systematically and commonly.

Real reform will require a change in the culture that prevails at the agencies. A system of rewards and punishments that responds to *both* Type 1 and Type 2 errors is needed. Government regulators have learned to avoid Type 1 errors at almost any cost because of the potent influence of the "stick" in a regulator's reward-punishment system. The agencies now need to create new sticks and carrots to redress the imbalance.

Some "carrots" exist in performance plan goals and the like, but these are, as the English say, weak as water. The clever regulator can always take refuge in the "inadequacy" or "inconclusiveness" of the data, and demand yet another study, another way of analyzing the data, another meeting of an advisory committee. Procrastination is a perennial problem of "pretesting" or "premarket" regulation.

Procrastination also has its benefits. The longer that product approvals take, and the more functions the agency assigns to itself, the larger the number of regulators needed. Larger budgets and larger empires to be managed are hardly disincentives.

The regulatory bureaucracy's current spectrum of incentives and disincentives encourages regulators to act like special interests. They tend to do what's best for themselves instead of what's best for patients or society as a whole. While most agency employees are dedicated and hardworking (and underpaid), they are forced to respond to the rewards and punishments of a system not of their own making. They are unlikely to rise above the level of expectations set by that system.

ESSAY 2
STRATEGIES FOR REFORM

The conundrum of what to do about government regulation of technology and its products requires a multifaceted solution. Industry trade associations need to define and better represent the long-term interests of their constituencies. Agencies need to implement specific reforms and also to change their culture of risk-aversion, in order to alter the current interplay of incentives and disincentives. The system needs to be better "gamed," in order to make self-interest constructive, rather than detrimental. Society's less visible stakeholders—patient, consumer and environmental groups—need to participate in the policymaking process and demand better.

THE QUEST FOR RISK-BASED REGULATION

The uses of the new biotechnology in "contained" laboratories, pilot plants, greenhouses and production facilities have engendered little controversy. The *NIH Guidelines for Research Involving Recombinant DNA Research* have exempted from oversight virtually all laboratory experiments, which has allowed organisms of low risk to be handled under minimal containment conditions. These conditions permit large numbers of living organisms to be present in the workplace and even to be released from the laboratory.[1] Despite extensive work in thousands of laboratories throughout the United States with millions of individual genetic clones, there has been no report of these incidental releases causing a human illness or any injury to the environment.

As discussed extensively in this volume, government regulation often has discriminated against the testing of rDNA-modified organisms in the environment. The key issue has centered on the scope of regulation—in other words, what experiments and products should fall into the regulatory net. Many negligible-risk experiments have been subjected to extreme regulatory scrutiny and lengthy delays solely because rDNA techniques were employed, even when the genetic change was completely characterized and benign and the organism demonstrably innocuous. As discussed in chapters 3 and 4, the impacts have been substantial. Investigators have shied away from areas of research that require field trials of recombinant organisms;[2] companies have avoided the newest, most precise and powerful techniques in order to manage R&D costs[3] and investors have avoided companies whose recombinant DNA-derived products became caught up in the public controversy and new regulation.[4]

Government agencies have variously regulated new biotechnology products using either previously existing regimes (FDA, until recently) or crafted new ones (EPA, USDA and NIH). Whether regulatory strategies are new or old, certain cardinal principles should apply. First, triggers to regulation—the criteria that determine which products and experiments warrant regulation—must be scientifically defensible. Second, the degree of oversight and compliance burdens must be commensurate with scientifically measurable risk. Some have contended that this may be obvious in theory but difficult to achieve in practice. Skeptics or critics of risk-based oversight contend that if we knew *a priori* which experiments and products were risky, agencies could just perform "armchair" risk assessment and exempt those proposals that pose negligible risk. Both assertions are weak.

The United States and other nations have devised other regulatory nets based on assumptions about the magnitude or the distribution of risk. For example, we require permits for field trials with certain organisms known or considered to be plant pests, whereas we exempt similar organisms based on a knowledgeable assessment of predicted risk. The validity of these assumptions determines the integrity of the regulatory scheme; without them, we might as well flip a coin or exempt field trials performed on certain days of the week.

Consistent with this regulatory philosophy, the federal government's 1986 *Coordinated Framework for the Regulation of Biotechnology* attempted, at least on paper, to focus oversight and regulatory triggers on the *characteristics* of products and on their intended use, rather than on the *processes* used for genetic manipulation.[5]

In spite of the Coordinated Framework's clearly-stated goals, the USDA and EPA have created oversight regimes for tests in the environment that conflicted with them. The agencies should have benefited from two landmark documents produced by the National Academy of Sciences[6] and the National Research Council.[7] They did not, however, and a number of U.S., foreign and international regulatory proposals have been based on process (chapters 3 and 4). For the most part, these proposals capture all rDNA-manipulated organisms. Sometimes the review requirement is limited to those which manifest phenotypes that "do not exist in nature," according to the rationale that such organisms are "unfamiliar," and by extension, potentially high risk. These proposals focus implicitly or explicitly on a process-determined definition of "familiar," an approach that seems to be derived from the prosaic meaning of the word, "accepted, accustomed, well-known." "Familiarity" is inappropriately equated with safety. Demonstrating the wrongheadedness and circularity of this approach, organisms are considered "familiar" solely because they are "natural" or have been created by older, more "familiar" genetic manipulation techniques. No matter how pathogenic, invasive or otherwise hazardous these "familiar" or "natural" organisms may be, they are intentionally exempted from the regulatory net.

A risk-based algorithm for field trials

As discussed extensively above, rDNA-manipulated organisms can be regulated in the same manner as other organisms. That is, according to intended use (such as vaccines, pesticides or food additives) or to intrinsic risk (a function of characteristics such as pathogenicity, toxigenicity and invasiveness). It is ironic that while various regulatory agencies in many nations have been struggling to make technique-based regulatory schemes plausible, risk-based alternatives have been widely available—and working.

Field trials of organisms that pose significant risk warrant biosafety oversight and appropriate precautions. In 1995, several colleagues and I refined an earlier biosafety algorithm that was mentioned in chapter 3. Like its predecessor, the more recent version of the algorithm[8] is scientifically defensible and risk-based. The basis of the algorithm is the tabulation of organisms into risk categories. The algorithm accommodates any organism, whether naturally occurring or genetically modified by old or new methods. It can provide the foundation for a cost-effective oversight system. It is adaptable to the resources and needs of different forms of oversight and regulatory mechanisms, whether they are implemented by governments or by other institutions.

First, in order to ascertain the degree of oversight appropriate to a wildtype, unmodified or parental organism in a field trial, a researcher would determine the "preliminary biosafety level," based on lists that stratify or categorize organisms according to risk. This tabulation would be based on scientific knowledge and experience as compiled by experts. A number of factors would determine this overall level of safety concern, including pest/pathogen status, ability to establish, location of centers of origin and dynamics of pollination (for plants), other ecological relationships and potential for monitoring and control.

Thus, the lists would provide an indication of the intrinsic level of risk of the organism, ranging from, say, Level 1 (lowest safety concern) to Level 5 (greatest safety concern). An important factor in stratifying plants according to potential risk, for example, would be the presence in a geographic area of cross-hybridizing relatives of the plant to be tested.[9] The proximity of a relative does not, however, alone confirm a risk. For example, there is limited gene flow from maize to nearby teosinte (and vice versa). Even when such gene flow occurs, it appears neither to be detrimental to the teosintes nor to change their basic nature as distinctive wild races and species.[10] Thus, the presence of teosinte near a field trial of maize does not alter the assessment of maize as posing negligible risk (category 1). By contrast, distinct varieties of oilseed rape (*Brassica napus,* or canola) with widely differing concentrations of erucic acid (and intended for different applications) should be kept segregated to avoid outcrossing between varieties. For example, high-erucic acid canola might be classified as category 1 in regions where that variety of the plant is grown but perhaps category 4 where low-erucic acid canola is grown.

This approach is analogous to that used for categorizing microorganisms by the U.S. Centers of Disease Control (CDC) and the National Institutes of Health (NIH) to establish laboratory safety standards for the handling of pathogens,[11] and to one proposed (but never implemented) by the USDA.[12] Foreign countries' regulatory approaches that employ inclusive lists of regulated articles such as plant pests or animal pathogens operate within similar principles.[13]

As a practical matter, it would be difficult for experts to stratify every organism in every geographic region according to risk. However, categorizing, say, a hundred of the major crop plants that are the likeliest candidates for field trials would be a feasible and useful beginning. Additional panels of experts could then address subsets of fish, terrestrial animals, microorganisms, insects and other groups.

Next, the preliminary categorization would be subject to reconsideration and adjustment in light of any new traits introduced, independent of the technique used for modification. The biosafety level could be adjusted up or down—on the basis of a major change in evolutionary fitness, pathogenicity, toxigenicity or invasiveness. Adjustments to higher categories might include, for example, field trials with pathogens manifesting new multiple antibiotic resistance or increased host range. Adjustments to lower categories could include a plant with decreased pollen production or a toxigenic microorganism after the complete deletion of its major toxin gene(s). Given the kinds of changes currently being made with molecular genetic techniques,[14] reclassification to higher biosafety levels would rarely be indicated.

Using the algorithm to determine degree of oversight

For regulators, especially those in the developing world, a crucial "first cut" is the designation of organisms that are considered to be of negligible risk (Level 1) or low risk (Level 2). This is important because, arguably, field trials in these lowest risk categories can be exempt from case by case review and managed using standard research practices appropriate to the test organism. By contrast, field trials with organisms in the highest risk categories should automatically require biosafety evaluation. (It is worth noting that this graduated approach to regulated and nonregulated field trials is both more scientific and more risk-based than either the pre- or post-rDNA regulatory regimes of EPA[15] or USDA;[16] see chapter 3.)

In theory, the degree of oversight of proposed field trials can vary widely between *exempt* (that is, subject to only the usual standard of practice for an agricultural experiment with that organism) and *prior approval required* (that is, by a national, regional or international agency), with various levels in between. However, we proposed only three levels of oversight: exempt, notification (to a local or international agency), or prior approval required.

In our scheme, the degree of required oversight can take into consideration not only the perceived level of risk but also other factors such as the available regulatory resources and the financial and manpower burden that regulation exacts from researchers and the government. Within the algorithm, a national or other policymaking authority could choose to apply regulatory strictures more stringently (tending toward more prior approval) or less stringently (tending toward more risk categories being exempt or requiring only notification). Of paramount importance, however, is that *the algorithm ensures that the overall approach is always within a scientifically sound context and that the degree of oversight is commensurate with risk.*

Many regulatory authorities would likely require case-by-case review for organisms in categories 4 and 5, exempt experiments in categories 1 and 2, and require a simple notification (describing the organism to be tested, the site, the risk management measures and so forth) for category 3. Within this internally-consistent scheme, other permutations are possible that would be chosen to meet regional preferences and needs.

This algorithm is very flexible and applicable to any organism. It meets the basic requirements of a biosafety regime—it is risk-based, scientifically defensible and focused on the characteristics of the test organisms and the environment of the field trial. The algorithm is highly adaptable; it can be incorporated into existing regulatory regimes in industrialized countries or be used by nations that currently lack such mechanisms. Moreover, it can offer adequate safety precautions to protect the public health and the environment from significant risk, coupled with the cost-effectiveness demanded by limited government resources.

Reform and the Agencies

The EPA

As discussed in chapter 3, EPA's biotechnology policies make neither scientific nor economic sense. They would require governmental review and high compliance costs for largely innocuous organisms, while exempting field trials of "naturally occurring" organisms and organisms manipulated by techniques other than rDNA that could foul waterways or pose other environmental risks.

It is both dismaying and ironic that the EPA's expenditure of hundreds of thousands of staff-hours in order to craft and implement new policies, over more than a decade, was unnecessary. Had agency officials simply applied the prevailing scientific consensus to their FIFRA and TSCA proposals, they would have concluded that no changes were needed from policies and requirements that were in effect before the advent of new biotechnology (just as FDA did). Products of the new rDNA technology would have been held to the same standards for testing and marketing as similar products used for similar purposes. Had regulators, the application of the new biotechnology to fields such as bioremediation, mining, oil recovery and pest control, that path now would be more advanced.

Little magic is needed to improve EPA's sorry state. The regulation of biotechnology products would benefit immediately from the application of the risk-based algorithm described above to the FIFRA and TSCA regulations. This could be easily accomplished.

It is unlikely that there is either interest or willingness to do so at EPA, however. The agency's regulatory infrastructure and personnel need a thorough, unbiased, nonpartisan, extramural review—and the recommendations should, for once, be implemented. The agency needs to incorporate avoidance of Type 2 errors into employees' performance plans and reviews, and would benefit from the kind of ombudsman panel described below that has the power to discipline agency officials for flawed policies or decisions. EPA needs to introduce accountability for regulators.

Equally important, we need national leaders committed to a strong role for science in public policy, including a role that includes comparative risk assessment in federal spending priorities and a "marketplace of ideas" in policy formulation. We need a knowledgeable, tough and competent EPA Administrator who will clean house and work with the Inspector General to deal severely with the kind of chicanery described above and in chapter 3. None of this is likely to happen during the Clinton-Gore administration. We can expect only more of the same.

The FDA

For the regulation of biotechnology, the FDA adopted a scientific paradigm early on that avoided discriminating against biotechnology products. Yet, the FDA still suffers from some of the same systemic problems as the EPA and USDA—distorted bureaucratic incentives and disincentives, ill-conceived policies and mismanagement. The FDA, which oversees more than $1 trillion worth of products annually, badly needs regulatory reform.

As discussed in chapter 3, the fear of comprehensive congressionally-mandated FDA reform stimulated the Clinton administration to announce a series of so-called "reforms" during 1995 and 1996. That Clinton administration officials are not serious about FDA reform, however, is clear from what was *not* included in the announced changes. They chose not to roll back recently instituted policies such as: broad new guidelines that require certain proportions of women and minorities in all federally funded clinical research, thereby slowing and increasing the costs of clinical trials, and encouraging companies to do the research abroad (these requirements were actually promulgated by NIH, as mandated by the 103rd Congress); new rules on the reporting of drug side effects; and restrictions on "promotional activities" such as informing physicians about peer-reviewed research findings and convening focus groups.[17]

The administration's changes seem more intended to impress the FDA's critics with a lengthy laundry list of "accomplishments" than to implement genuine change. The minimal benefits will accrue principally to larger, established companies that already have products on the market. Entrepreneurial companies, whose products are primarily in early developmental stages, are left out.

The most serious flaw in the administration's proposals—one likely to vitiate much of their impact—is that the reforms are to be implemented by the FDA itself. Experience should have taught the futility of an agency being directed to reform itself. Similar reforms, instigated by President Bush's Council on Competitiveness, were announced by the FDA in 1991.[18] These were conservative but potentially consequential. They included such changes as outside organizations performing reviews (under contract to FDA); expanded use of extramural advisory committees; an expanded role for Institutional Review Boards; a more flexible interpretation of the efficacy standard; U.S. recognition of foreign approvals (that is, reciprocity) and various management improvements. An important element of many of these reforms was *structural* change—which would actually diminish the scope of FDA's discretion or jurisdiction—rather than mere managerial tinkering. To no one's surprise, the agency studied them, literally, to death. As an FDA official at the time, I recall agency officials' amusement at the prospect of their reforming themselves. And that was during a presidential administration that really did care about streamlining regulation.

Minimalist but effective reform

No one who has worked at a regulatory agency is likely to be terribly optimistic about the prospect of dramatically improving the "culture" of rank-and-file agency employees (although there are suggestions for accomplishing this, below). Therefore, a "first practical principle" of reform should be to strive also for *structural* changes which reduce the agency's discretion and influence. This can be accomplished by exempting entirely certain regulated activities or products, or by transferring regulatory functions to non-governmental entities where the professional culture and the incentives and disincentives that shape behavior are more propitious.

Applying these principles, the Congress could, at a stroke, achieve real reform. A few narrow but critical amendments to the FDA's enabling statutes would remove certain functions from the governmental monopoly and reduce the agency's opportunities for mischief:

- Exempt from FDA jurisdiction early small-scale clinical trials, which are overseen already by research institutions' Institutional Review Boards;
- Reduce or eliminate FDA control over drug advertising and promotion;
- Require the FDA to recognize drug approvals by comparable regulatory apparatuses abroad (e.g., the U.K., Canada and the European Medicines Evaluation Agency);
- Direct the FDA to certify private-sector entities to perform reviews of clinical trials;
- Eliminate the FDA from the review of exports of experimental drugs and medical devices; and
- Establish statutory "hammers" (waiting periods after which approval is automatic), which would compel the agency to meet mandated time limits for product review.

These genuine reforms would offer no less safety to consumers but would confer several benefits. They would lessen the regulatory load—and the costs—of pharmaceutical development, permit the FDA to focus on essential functions and provide physicians and patients more (and less expensive) therapeutic alternatives.

While these reforms would improve the efficiency of drug regulation and do no harm to consumers in the process, none of them addresses the fundamental problem of the distorted incentives, disincentives, rewards and punishments that influence federal regulators' behavior.

Avoidance of Type 1 errors implicitly is already built into reviewers' and managers' performance plans. An employee whose actions compel his superiors to defend his mistaken approval of a hazardous product will suffer during his annual performance review. The system should similarly foster aversion to Type 2 errors. For all FDA (and EPA) employees involved in product evaluation or compliance, therefore, performance plans and employee annual reviews should be required to give equivalent weight to Type 1 and Type 2 errors. (Similar performance plan elements, such as meeting affirmative action goals, are currently applied at least as widely at federal agencies.) This new element would require no more than the kind of cost/benefit critical judgments that managers are routinely supposed to perform.

Another remedy is an ombudsman panel that evaluates agency actions and disciplines misbehavior. Cases could be submitted to the panel by drug companies, associations, patient groups or others.

My own experience as an FDA reviewer and manager suggests a number of cases appropriate for the ombudsman. During the 1980s, despite a demonstration that a certain proven anticancer agent could shrink the malignant Kaposi's sarcoma lesions found in AIDS, the FDA would not consider approval for that use. The agency said, in effect, that such an effect was merely cosmetic, and that the drug's sponsor needed to show improvement of a "meaningful" endpoint, such as patient survival. Even though that decision was eventually reversed, the officials who were responsible for it should be held accountable.

An ombudsman panel must have several characteristics: (1) an organizational location outside the agency, to provide arm's length from FDA officials, including the Commissioner (the Department of Health and Human Services' Office of Inspector General might be an appropriate location); (2) access to a wide spectrum of scientific, medical and regulatory expertise, either via a large membership or ad hoc experts as necessary; and (3) authority to recommend disciplinary sanctions, ranging from censure to forfeited pay and bonuses or demotion, depending on the egregiousness and impact of the decision.

The actions of the ombudsman panel would redress, in part, the agency's tendency to guard against approving a harmful product even at the expense of erecting huge economic barriers to R&D and marketing. This innovative mechanism could help to balance regulators' incentives and disincentives. More fundamentally, it could be applied to other regulatory agencies. It is a conservative governmental mechanism for correcting the entrenched bias toward eliminating product risk regardless of the cost of lost benefits.

More sweeping reform: the Progress and Freedom Foundation proposal and HR 3199

In February 1996, the Washington D.C.-based Progress and Freedom Foundation (PFF) published a comprehensive analysis of the FDA, along with proposals for the reform of drug and medical device regulation.[19] (I was one of the authors.) Their solution to the systemic problems is to turn over much of the evaluation of drugs to nongovernmental entities—a recommendation that has been made repeatedly by blue-ribbon expert groups convened to evaluate the drug-approval system.

The PFF proposal would retain the basic requirements that a drug be proved safe and effective before marketing. It mandates, however, that "Drug Certification Bodies" (DCBs) supplant the FDA in two important ways: overseeing drug companies' clinical testing and, after testing is completed, performing the primary review of the data that supports marketing.

The DCBs can be private- or public-sector organizations (universities, for example), profit-making or nonprofit. They would be subject to FDA certification and auditing, and staffed with "experts qualified by scientific training and experience," as required by statute.

The drug sponsor would submit the request for approval to market a drug to the DCB rather than to the FDA. After evaluating it, the DCB would submit the results (if favorable) to the FDA. The FDA Commissioner would have a designated period of time in which to accept or deny the DCB's recommendation for approval. In the event that the FDA denied approval, an appeal mechanism would be available to the drug sponsor.

The PFF proposal is derived both from first principles and careful study of three decades' experience with the FDA's drug regulation. As discussed in this essay and in chapter 3, a fundamental problem at the FDA is the agency's tendency to slow the approval process to avoid even the remotest possibility of approving a product that might be harmful. The costs are high and the profitability threshold has risen, as the price to bring a drug to market has rocketed to around $500 million.[20] Many important therapies have been delayed or abandoned. A contributing factor is the FDA's absolute regulatory "monopoly." For drug approval, the feds are the only game in town.

The PFF plan redresses this problem ingeniously by introducing the element of competition into the drug review process. DCBs would compete for clients, and thus, the system would favor those that offer the greatest expertise, the best service and a reputation for integrity. In contrast to the FDA, where bureaucratic incentives and disincentives encourage a "go slow" mind set, competition would encourage DCBs to devise new, more innovative and efficient ways for their clients to demonstrate product safety and efficacy.

DCBs would be discouraged in several ways from the temptation of a "quick and dirty" but lucrative review: the threat of losing FDA certification; the need for FDA's final sign-off on the approval and the risk of legal liability, should the product ultimately cause harm after an inadequate review. This balance is not unlike that confronted by Underwriters' Laboratories, which certifies that electrical equipment meets certain safety standards. It also closely resembles the system of medical device regulation in the European Union.

It is remarkable that within two months of its publication major elements of the PFF proposal found their way into proposed legislation, HR 3199, the House of Representatives' Drugs and Biological Products Reform Act of 1996.

The bill addresses in several ways the ponderousness and length of drug testing and evaluation. It clarifies that the voluminous raw data from clinical trials—often running to hundreds of thousands, sometimes millions of pages—will not always be required by the FDA. Condensed, tabulated or summarized data often will be adequate. Agency reviewers would have access to additional material if it were requested by supervisory FDA officials.

HR 3199 also emphasizes that the demonstration of efficacy of a new drug may be based on even a single "well-controlled" trial, if the statistical analysis and reviewers' judgment support it.

This change removes ambiguity in the language of the existing statute. In addition, taking aim at the FDA's tendency to require double-blind trials (where both patient and physician are ignorant whether the treatment is with active drug or placebo) under all circumstances, the bill mandates that clinical trial design should be "appropriate to the intended use of the drug and the disease."

The legislation establishes a new, more liberal efficacy standard for drugs intended to treat "serious or life-threatening" conditions. Like the current standard for AIDS drugs that has spurred rapid approvals, experts would have to conclude that "there is a reasonable likelihood that the drug will be effective in a significant number of patients and that the risk from the drug is no greater than the risk from the condition." This is just common sense. It is also humane: it would permit other patients the same treatment currently reserved for those with AIDS.

The bill would change in important ways FDA's censorship of scientific and medical information (chapter 3). The FDA currently prohibits drug companies from distributing textbooks and articles to health professionals if they contain information about not-yet-approved uses of drugs—even though these constitute almost half of all physicians' prescriptions, about 70% of all cancer chemotherapy and 90% of drugs used in pediatrics.[21] The bill would permit the legitimate dissemination of information via textbooks and articles from peer-reviewed journals.

Equally important because it goes to the basic issue of how new uses are sanctioned for an already-approved drug, HR 3199 would permit retrospective evidence from clinical research (instead of expensive and time-consuming prospective studies) as an alternate basis for approving additional uses.

The bill includes some modest incentives for the FDA to improve its performance, such as mandatory annual reports to the Congress that summarize the agency's success in meeting goals and statutory deadlines and that compare the FDA's performance with its foreign counterparts. To ensure that the FDA's attempts at international harmonization are sensible, the bill would require congressional notification before the FDA enters into international agreements.

The most significant changes wrought by the legislation—similar to but less sweeping than the PFF proposal—address the FDA's monopoly over the drug approval process. These provisions would provide a partial answer to systemic problems in the regulation of drug development, by turning over part of the evaluation of drugs to nongovernmental entities.

These proposals to "privatize" certain regulatory activities have been savaged by defenders of big government. There is, however, nothing sacrosanct about a government regulatory monopoly that offers manufacturers no alternative route to the review and certification of products. Moreover, regulation that assures public safety is not a binary choice—that is, either government or private-sector. That is illustrated by the bill's establishment of nongovernmental alternatives to some FDA oversight. Drug sponsors could opt to have their products reviewed by nongovernmental organizations that could be private- or public-sector (universities, for example), profit-making or nonprofit. These organizations would be subject to FDA accreditation and auditing. Strict requirements backed by civil and criminal sanctions would assure the confidentiality of data and the management of potential conflicts of interest.

This new approach closely resembles regulatory apparatuses already operating elsewhere, except that the sponsor could choose to have the FDA perform the review and in all cases the agency would retain the responsibility for final sign-off.

HR 3199 omits any concrete provisions for reciprocity, which would hasten the approval of a drug in the United States after its sanction by a major foreign regulatory authority. Reciprocity could be achieved, for example, simply by giving the FDA a finite period of time (say, 60 days) from the date of a UK or EU approval to show cause why a product should not be approved. In the absence of such evidence from the FDA (which carries the burden of proof), the drug would be approved automatically.

Overall, the House Commerce Committee's solutions to FDA reform are almost too good to be true. They are logical, carefully targeted, and bipartisan and—most important—favor the public interest. They would get drugs to patients who need them, faster and cheaper. Legislation should balance momentous social and economic issues, legal precedents and the public interest. HR 3199, which strongly reflects the influence of House Commerce Committee majority counsel John Cohrssen (and is a kind of reprise of his earlier attempts to reform FDA while at the Bush administration's Council on Competitiveness), is a stunning example.

ESSAY 3
BUILDING A CONSTITUENCY FOR SCIENCE-BASED POLICIES

Habituation, the gradual adaptation to a specific stimulus or to the environment, is a biological phenomenon that may also be said to apply to responses to political influences. Irrational and burdensome public policies can become so much a part of the landscape that their victims—consumers, businesses and research institutions—no longer experience the appropriate rush of self-righteous anger and push for reform. The worst becomes the norm.

Habituation can be observed in the unresponsive attitudes of academic and industrial scientists towards the excessive regulation of agricultural and environmental biotechnology. Scientists should actively question the flawed paradigms underlying the regulations. Few do. Putting this another way, the members of the research community need to pursue their self-interest aggressively, as other groups do.

We have seen the behavior of industry trade associations toward ill-conceived or excessive biotechnology regulation (above and chapter 3). The expected chorus of indignation from individual agricultural biotechnology researchers and companies has been absent. Surprisingly, most have settled for clarity instead of reasonableness. They have settled for predictability, even though it is the predictability of delay and frustration. Often, rather than working to make regulation more reasonable investigators have changed the direction of their research to avoid regulatory strictures.

Exceptions include a handful of individual scientists, the University of California's Systemwide Biotechnology Program, and a few professional societies that have called for the rationalization of regulation in editorials and letters to government agencies.[1] Another notable exception is the August 1996 report from no fewer than *eleven* scientific societies that excoriates EPA's plant-pesticide regulatory proposal.[2] This development illustrates that the threshold for definitive action is high—the plant-pesticide proposal is only the latest in a decade-long string of scientifically bankrupt EPA policies—but when it is reached, the scientific community can be mobilized.

Federal regulatory policies have given rise to disincentives to R&D in various sectors of U.S. biotechnology, accompanied by the disenchantment of researchers and investors (chapters 2 and 3).

Milton Friedman diagnosed correctly that the "government is the problem" and that the problem is with the system, not with the people (although an exception may be EPA, where, as discussed in chapter 3 individuals appear to be culpable).[3] The self-interest of government officials often causes them to behave in a way that is inimical to the self-interest of the rest of us.

If those who are interested in the vitality of science, technology and innovation sit on their hands, nothing will change—except, perhaps, for the worse. The agencies seem unwilling and unable to reform themselves. The Clinton administration certainly is not serious about reforms.

Earlier in this chapter, I suggested ways that government regulation could be improved by structural and management changes, along with a risk-based algorithm for the oversight of field trials. But government left to its devices won't adopt these changes. It is past time that those outside government began to hold policy makers accountable and to exert pressure. But whence is this pressure to come?

I suggest six strategies for "progress"—defined as the integration into public policy of scientific, risk-based, regulatory approaches. First, scientists, as individuals, must do more of what physicist and writer Freeman Dyson, paleontologist Stephen Jay Gould and the late microbiologist Bernard Davis have done in their articles and books: they have participated in the dialogue on public policy issues. As scientists, they have made unique contributions, especially when exposing nonscientific arguments. Whether one agrees or disagrees with their arguments, their contributions are invaluable. This kind of involvement should occur in every possible forum, including scientific and "popular" articles, communication with the news media, and—especially—scientific advisory panels at government agencies.

This strategy is not, however, without its risks. No matter how brilliant a scientist may be in his specialty, acuity in public policy requires a different perspective. I was reminded of this by a 1996 *Science* editorial by yeast geneticist Gerry Fink.[4] Fink recounted how in 1977, National Science Foundation administrator Herman Lewis found a way to circumvent the NIH recombinant DNA guidelines' prohibition on doing certain cloning experiments in yeast; this permitted Fink and his co-workers to do the experiments, literally years before they would have been able otherwise.

Their experiments accelerated research leading ultimately to the development of a much improved, second-generation hepatitis B vaccine (of which I was one of the FDA reviewers; see discussion above). But in his editorial Fink neglected the critical point—that other U.S. researchers who lacked a governmental good Samaritan were stymied for years by regressive, unnecessarily restrictive federal regulatory policies. Because Lewis was such a notable exception to the rule, the story should have been about bad science making bad policy, and the real-world impacts of bad policy. But Fink lacked a perspective on the wider dimensions of the NIH policy.

The second strategy pertains to science in its institutional forms—the professional associations, faculties, academies and journals. These institutions should explore and elucidate the controversies over public policy and seek to elevate the level of discourse on them. Scientific societies can, for example, help to create and promote a broader policy perspective by building public policy symposia into national and international conferences. And to return to the example of the Fink editorial, the editors at *Science* could have steered the piece to the more didactic and broader theme.

Reporters—and by extension, their bosses, the editors—have tremendous power to illuminate public policy issues that have a scientific component. Too often, in the interest of "balance," all of the views on an issue are presented as though they were of equal weight or value—even after the dialogue has progressed to a point where some views have already been discredited. The Flat Earth Society should not receive the same attention and credence as the geography department at Berkeley, when the former eschews empirical evidence. It should be a "given" that advocates of "creation biology" are less deserving than Harvard cosmologist and paleontologist Stephen Jay Gould of having their views sought on evolutionary issues. Likewise, on many biotechnology issues, the media need to—but often do not—reflect the current status of the "debate." Unfortunately, by manufacturing or exaggerating a controversy they often get a "better" story.

Third, companies and trade associations should consistently take the long view of regulation, one dictated by scientific and free-market economic principles, and resist the temptation of flawed short-term fixes. There is no sound reason for the U.S.

biotechnology industry to support or encourage policies like the USDA's Plant Pest Act regulations, the EPA's technique-based proposals or the FDA's proposal to require the submission of data for all new biotechnology foods.

In the long-run, commercial interests will benefit from the predictability and logic of science-based policies—and from a robust academic research enterprise. Productivity is squandered by strategies that are anticompetitive, that make experiments more expensive and difficult, and that force researchers to do mountains of unnecessary paperwork instead of experiments.

Fourth, those who are not directly involved in science but who are important stakeholders in the ultimate applications of science and technology—venture capitalists, consumer groups, patients' groups—should commission experts to help them both to discern and advocate rationality in governmental oversight of R&D and product marketing.

Fifth, government officials need dramatic behavioral modification. Consider the old story about the city slicker who sees a farmer whacking his mule with a two-by-four. The city slicker asks the farmer what he wants the animal to do. "Nothing, yet," the farmer replies, "I'm just trying to get his attention." As in the story, American taxpayers need to get bigger sticks. Negative reinforcement, appropriately applied, can redress some of the existing asymmetry between Type 1 and Type 2 errors. For example, independent ombudsman panels that have the authority to discipline individuals at regulatory agencies for egregious errors would introduce accountability into regulators' policy-making and decision-making.

Sixth, the government/nongovernment balance of influence should be shifted. There is nothing sacrosanct about a government monopoly over regulation.

There are a variety of alternative institutional arrangements that could serve to oversee the premarket approval and monitoring of products like drugs, medical devices, food additives and pesticides. As economist Robert Tollison has said of pharmaceutical regulation:

> There are a variety of alternative institutional arrangements which could be called upon to regulate and monitor the safety and effectiveness of the nation's supply of new drugs and medical devices. These range from the present system, which can be classified as a government monopoly

> on drug and device certification, to a free market system in which government would have no regulatory role at all. In between these two extremes are a number of institutional alternatives which combine more or less government oversight with more or less private involvement. There are costs and benefits of each system, and it is through a careful consideration of these costs and benefits that one can proceed to choose a system that provides the most net benefits to medical consumers.[5]

As I have argued throughout this book, the costs of the present system often outweigh its benefits; I have identified several examples where the balance is overwhelmingly negative. It is past time for the regulatory pendulum to swing away from government monopoly to a part of the arc that emphasizes the innovation and efficiency favored by nongovernmental mechanisms.

The remedies I have suggested seem the only routes to altering, even in this small realm and in a limited way, the validity of historian Barbara Tuchman's observation that "[m]ankind, it seems, makes a poorer performance of government than of almost any other human activity."[6]

REFERENCES

Essay 1: Onerous Regulatory Policies: Blame Industry!

1. Bovard J. First Step to an FDA Cure: Dump Kessler. The Wall Street Journal, December 8, 1994; A16. See also Miller HI. When Politics Drives Science. The Los Angeles Times December 12, 1994. Miller HI. Dr. Kessler's regulatory obsessions. The Washington Times December 15, 1994; A21.
2. Bovard, J. Ibid.
3. Crooke ST. Comprehensive Reform of the New Drug Regulatory Process. Bio/Technology 1994; 13:25-29.
4. Sears G, van Beek J, Golder G. Improving Canadian Biotechnology Regulation—A Study of the U.S. Experience. 1995 consultants' report. I was a consultant on this report but not one of its authors.
5. Anon. Statement of policy: foods derived from new plant varieties. Federal Register 1992; 57:22984-23005.
6. Feldbaum, CF. Letter from BIO to Congressman Don Young, July 17, 1995.
7. Nettleton J. Personal communication.

8. Anon. Unified regulatory agenda. Federal Register 1995; 60:23291.
9. Hoyle R. Biosafety protocol draft spooks U.S. biotechnology officials. Nature Biotechnology 1996; 14:803. See also Kaiser J, U.S. Frets Over Global Biosafety Rules. Science 1996; 273:299.
10. Idem.
11. Helms RB. Preface to Drugs and Health: Economic Issues and Policy Objectives. Washington D.C.: American Enterprise Institute for Public Policy Research, 1981:xx-xxiii.

Essay 2: Reform Strategies

1. Lincoln DR, Fisher ES, Lambert D et al. Release and Containment of Microorganisms from Applied Genetics Activities, report submitted in fulfillment of EPA Grant No. R-808317-01, 1983.
2. Ratner M. BSCC addresses scope of oversight. Bio/technology 1990; 8:196-8. See also UW researchers stymied by genetic test limits. The Capital Times (Madison, WI) March 16, 1988:31.
3. Naj AK. Clouds Gather Over the Biotechnology Industry, The Wall Street Journal 1989; 11.
4. Miller HI. Governmental regulation of the products of the new biotechnology: A U.S. perspective. In: Proceedings of Trends in Biotechnology: An International Conference Organized by the Swedish Council for Forestry and Agricultural Research and the Swedish Recombinant DNA Advisory Committee. Stockholm: AB Boktryck HBG, 1990.
5. Anon. Coordinated framework for regulation of biotechnology. Federal Register 1986; 51:23302-93.
6. Anon. Introduction of Recombinant DNA-Engineered Organisms into the Environment: Key Issues, Washington D.C.: Council of the U.S. Academy of Sciences/National Academy Press, 1987.
7. Anon. Field Testing Genetically Modified Organisms: Framework for Decisions. Washington D.C.: U.S. National Research Council/ National Academy Press, 1989.
8. Miller HI, Altman DW, Barton JH et al. Biotechnology Oversight in Developing Countries: A Risk-Based Algorithm. Bio/Technology 1995; 13:955-59.
9. Anon. Field Testing Genetically Modified Organisms: Framework for Decisions. Ibid.
10. Idem.
11. Anon. Biosafety in Microbiological and Biomedical Laboratories, Centers for Disease Control/National Institutes of Health, U.S. Department of Health and Human Services. Washington D.C.: U.S. Government Printing Office, 1988.
12. Anon. Proposed guidelines for research involving the planned introduction into the environment of organisms with deliberately modified traits. Federal Register 1991; 56:4134.

13. Frommer W, Ager B, Archer L et al. Safe biotechnology: III. Safety precautions for handling microorganisms of different risk classes. Appl Microbiol Biotech 1989; 30:541.
14. Chrispeels MJ, Sadava DE. Plants, Genes and Agriculture. Boston: Jones and Bartlett, 1994:chapter 15.
15. Anon. Coordinated framework for regulation of biotechnology. Ibid.
16. Idem.
17. Miller, HI. Anti-Medicine Man. National Review 1995; 48-51.
18. Anon. Council on Competitiveness Fact Sheet: Improving the Nation's Drug Approval Process. Washington D.C.: The White House, November 13, 1991.
19. Epstein RA, Lenard TM, Miller HI et al. Advancing Medical Innovation. Washington D.C.: The Progress and Freedom Foundation, 1989.
20. The Boston Consulting Group analysis based on data of J. DiMasi et al, as quoted by the Office of Technology Assessment in Pharmaceutical R&D: Costs, Risks, and Rewards, Washington D.C., February 1993. Estimate is in pretax 1990 dollars.
21. Henderson DR. FDA censorship can be hazardous to your health. Policy Brief 158, St. Louis: Center for the Study of American Business, September 1995.

Essay 3: Building a Constituency for Science-Based Policies

1. Arntzen C. Regulation of Transgenic Plants. Science 1992; 257:1327. Also Huttner SL, Arntzen C, Beachy R et al. Revising Oversight of Genetically-Modified Plants. Bio/Technology 1992; 10:967-971; and Anon. ASBMB [American Society for Biochemistry and Molecular Biology] News Winter 1993; 2:1.
2. Anon. Appropriate Oversight for Plants with Inherited Traits for Resistance to Pests, 1996.
3. Friedman M. Why government is the problem. In: Hoover Institution Essays in Public Policy, Stanford: Hoover Institution Press, 1993.
4. Fink G. Bureaucrats Save Lives. Science 1996; 271:1213.
5. Tollison RD. Institutional alternatives for the regulation of drugs and medical devices. In: Advancing Medical Innovation. Washington D.C.: The Progress & Freedom Foundation, 1996:18.
6. Tuchman BW. The March of Folly: From Troy to Vietnam. Boston: Knopf, 1984:4.

INDEX